Electronics Cookbook

*Practical Electronic Recipes
with Arduino and Raspberry Pi*

Simon Monk

Beijing · Boston · Farnham · Sebastopol · Tokyo

Electronics Cookbook

by Simon Monk

Copyright © 2017 Simon Monk. All rights reserved.

Printed in the United States of America.

Published by O'Reilly Media, Inc., 1005 Gravenstein Highway North, Sebastopol, CA 95472.

O'Reilly books may be purchased for educational, business, or sales promotional use. Online editions are also available for most titles (*http://oreilly.com/safari*). For more information, contact our corporate/institutional sales department: 800-998-9938 or *corporate@oreilly.com*.

Editors: Susan Conant and Jeff Bleiel
Production Editor: Colleen Lobner
Copyeditor: Christina Edwards
Proofreader: Kim Cofer

Indexer: Judy McConville
Interior Designer: David Futato
Cover Designer: Karen Montgomery
Illustrator: Rebecca Demarest

April 2017:   First Edition

Revision History for the First Edition
2017-03-29:   First Release

See *http://oreilly.com/catalog/errata.csp?isbn=9781491953402* for release details.

978-1-491-95340-2

[LSI]

Table of Contents

Preface

Traditional wisdom requires people using electronics to have at least an EE degree before they can do anything useful, but in this book the whole subject of electronics is given the highly respected O'Reilly Cookbook treatment and is broken down into recipes. These recipes make it possible for the reader to access the book at random, following the recipe that solves their problem and learning as much or as little about the theory as they are comfortable with.

While it is impossible to cover in one volume everything in a complex and wide-ranging subject like electronics, I have tried to select recipes that seem to come up most frequently when I talk to other makers, hobbyists, and inventors.

Who Should Read This Book

If you are into electronics or want to get into electronics, then this is the book that will help you get more from your hobby. The book is full of built-and-tested recipes that you can trust to do just what you need them to do, no matter what your level of expertise.

If you are new to electronics then this book will serve as a guide to get you started; if you are an experienced electronics maker, it will act as a useful reference.

Why I Wrote This Book

This book has been gestating for a while. I believe that the original concept came from no less a person than Tim O'Reilly himself. The idea was to fill the gap in the market between books like the *Arduino Cookbook* and the *Raspberry Pi Cookbook* and heavyweight electronics textbooks.

In other words, to cover more of the fundamentals of electronics and topics peripheral to the use of microcontrollers that often get neglected, except in heavyweight electronic tomes. Topics such as how to construct various types of power supply, using the right transistor for switching, using analog and digital ICs, as well as how to construct projects and prototypes and use test equipment.

A Word on Electronics Today

Boards like the Arduino and Raspberry Pi have lured whole new generations of makers, hobbyists, and inventors into the world of electronics. Components and tools are now low cost and within the reach of more people than at any time in history. Hackspaces and Fab Labs have electronic workstations where you can use tools to realize your projects.

The free availability of information including detailed designs means that you can learn from and adapt other people's work for your own specific needs.

Many people who start with electronics as a hobby progress to formal education in electronic engineering, or just jump straight to product design as an inventor and entrepreneur. After all, if you have access to a computer and a few tools and components, you can build a working prototype of your great invention and then find someone to manufacture it for you, all financed with the help of crowdfunding. The barrier of entry to the electronics business is at an all-time low.

Navigating This Book

As a "cookbook" you can dive in and use any recipe, rather than read the book in order. Where you have a recipe that relies on some knowledge or skills from another recipe, there will be a link back to the prerequisite recipe.

The recipes are arranged in chapters, with Chapters 1 to 6 providing more fundamental recipes, some concerning theory but mostly about different types of component (your recipe ingredients). These chapters are:

- Chapter 1, *Theory*. As the title suggests, the recipes in this chapter provide you with the few theoretical concepts such as Ohm's Law and the power law you just can't avoid.
- Chapter 2, *Resistors*. These most common of electronic components are explained and recipes provided for some of their uses.
- Chapter 3, *Capacitors and Inductors*. Here you will find recipes explaining how these components work, how to identify them, and recipes for making use of them.
- Chapter 4, *Diodes*. In this chapter you will find recipes explaining diodes and uses for different types of diode including Zener diodes, photodiodes, and LEDs.
- Chapter 5, *Transistors and Integrated Circuits*. This chapter mostly contains fundamental recipes for using transistors and guides for using different types of

transistors in different settings. ICs (integrated circuits) are introduced, but you will find individual recipes for ICs scattered throughout the rest of the book.

- Chapter 6, *Switches and Relays*. The section ends with a look at these common but often overlooked components.

The next section of chapters looks at how the components introduced in the first section can be used together in various recipes covering pretty much anything electronic that you might like to design.

- Chapter 7, *Power Supplies*. Whatever your project, you are going to need to provide it with power. You will find recipes here for both traditional power supply designs as well as switched mode power supplies (SMPS) and more exotic high-voltage power supplies.
- Chapter 8, *Batteries*. This chapter contains recipes for selecting batteries and also practical circuits for charging batteries (including LiPo batteries) and automatic battery backup.
- Chapter 9, *Solar Power*. In this chapter, you will find recipes to help you power your projects using solar panels, including providing solar power to an Arduino and Raspberry Pi.
- Chapter 10, *Arduino and Raspberry Pi*. Most Maker projects now include the use of a computing element like an Arduino or Raspberry Pi. These boards are introduced along with some recipes for using them to control external electronics.
- Chapter 11, *Switching*. Not to be confused with "switches," this chapter provides recipes that show you how to use transistors, electromechanical relays, and solid-state relays to turn things on and off using an Arduino or Raspberry Pi.
- Chapter 12, *Sensors*. This chapter is packed with recipes for many different types of sensor and shows you how to use them with both Arduino and Raspberry Pi.
- Chapter 13, *Motors*. In this chapter, there are recipes for using different types of motors (DC, stepper, and servo) with both Arduino and Raspberry Pi. There are also recipes for controlling both the speed and direction of motors.
- Chapter 14, *LEDs and Displays*. In addition to recipes for controlling standard LEDs from an Arduino or Raspberry Pi, this chapter also has recipes for using high-power LEDs and various types of displays, including OLED graphical displays, addressable LED strips (NeoPixels), and LCD displays.
- Chapter 15, *Digital ICs*. This chapter contains recipes for using those digital ICs that are still useful in your projects in spite of the advent of microcontrollers.
- Chapter 16, *Analog*. In this chapter, you will find a collection of recipes for various useful analog designs from simple filtering to a range of oscillator and timer designs.
- Chapter 17, *Operational Amplifiers*. Continuing with the analog theme, this chapter provides recipes for using op-amps for various tasks from straightforward amplification to filter design, buffering, and comparators.

- Chapter 18, *Audio*. Here, you will find recipes for making sounds from an Arduino or Raspberry Pi as well as power amplifier designs (both analog and digital) and amplifying the signal from a microphone.
- Chapter 19, *Radio Frequency*. This chapter has some interesting recipes for FM transmitters and receivers as well as for sending packet data from one Arduino to another.

The final section of the book contains recipes for construction and the use of tools.

- Chapter 20, *Construction*. This chapter contains recipes for building "unsoldered" prototypes and for making those projects into a more permanent soldered form. It also provides recipes for soldering, both through-hole and surface-mount devices.
- Chapter 21, *Tools*. The use of bench power supplies, multimeters, oscilloscopes, and the use of simulations software are all described here in a series of recipes.

The book also includes appendices that list all the parts used in the book along with useful suppliers and provide pinouts for devices including the Arduino and Raspberry Pi.

Online Resources

There are many wonderful resources available for the electronics enthusiast.

If you are looking for project ideas then sites like Hackaday (*http://hackaday.com*) and Instructables (*http://instructables.com*) are a great source of inspiration.

When it comes to getting help with a project, you will often get great advice from the many experienced and knowledgable people that hang out on the following forums. Remember to search the forum before asking your question, in case it has come up before (usually it has) and always explain your question clearly, or "experts" can get impatient with you.

- *http://forum.arduino.cc*
- *https://www.raspberrypi.org/forums*
- *http://www.eevblog.com/forum*
- *http://electronics.stackexchange.com*

Conventions Used in This Book

The following typographical conventions are used in this book:

Italic

    Indicates new terms, URLs, email addresses, filenames, and file extensions.

`Constant width`

> Used for program listings, as well as within paragraphs to refer to program elements such as variable or function names, databases, data types, environment variables, statements, and keywords.

`Constant width bold`

> Shows commands or other text that should be typed literally by the user.

`Constant width italic`

> Shows text that should be replaced with user-supplied values or by values determined by context.

 This element signifies a tip or suggestion.

 This element indicates a warning or caution.

Using Code Examples

Supplemental material (code examples, exercises, etc.) is available for download at *https://github.com/simonmonk/electronics_cookbook*.

This book is here to help you get your job done. In general, if example code is offered with this book, you may use it in your programs and documentation. You do not need to contact us for permission unless you're reproducing a significant portion of the code. For example, writing a program that uses several chunks of code from this book does not require permission. Selling or distributing a CD-ROM of examples from O'Reilly books does require permission. Answering a question by citing this book and quoting example code does not require permission. Incorporating a significant amount of example code from this book into your product's documentation does require permission.

We appreciate, but do not require, attribution. An attribution usually includes the title, author, publisher, and ISBN. For example: "*Electronics Cookbook* by Simon Monk (O'Reilly). Copyright 2017 Simon Monk, 978-1-491-95340-2."

If you feel your use of code examples falls outside fair use or the permission given above, feel free to contact us at *permissions@oreilly.com*.

O'Reilly Safari

 *Safari* (formerly Safari Books Online) is a membership-based training and reference platform for enterprise, government, educators, and individuals.

Members have access to thousands of books, training videos, Learning Paths, interactive tutorials, and curated playlists from over 250 publishers, including O'Reilly Media, Harvard Business Review, Prentice Hall Professional, Addison-Wesley Professional, Microsoft Press, Sams, Que, Peachpit Press, Adobe, Focal Press, Cisco Press, John Wiley & Sons, Syngress, Morgan Kaufmann, IBM Redbooks, Packt, Adobe Press, FT Press, Apress, Manning, New Riders, McGraw-Hill, Jones & Bartlett, and Course Technology, among others.

For more information, please visit *http://oreilly.com/safari*.

How to Contact Us

Please address comments and questions concerning this book to the publisher:

O'Reilly Media, Inc.
1005 Gravenstein Highway North
Sebastopol, CA 95472
800-998-9938 (in the United States or Canada)
707-829-0515 (international or local)
707-829-0104 (fax)

We have a web page for this book, where we list errata, examples, and any additional information. You can access this page at *http://bit.ly/electronics-cookbook*.

To comment or ask technical questions about this book, send email to *bookquestions@oreilly.com*.

For more information about our books, courses, conferences, and news, see our website at *http://www.oreilly.com*.

Find us on Facebook: *http://facebook.com/oreilly*

Follow us on Twitter: *http://twitter.com/oreillymedia*

Watch us on YouTube: *http://www.youtube.com/oreillymedia*

Acknowledgments

Thanks to Duncan Amos, David Whale, and Mike Bassett for their technical reviews of the book and the many useful comments that they provided to help make this book as good as it could be.

I'd also like to thank Afroman (*http://afrotechmods.com/*) for permission to use his great FM transmitter design and the guys at Digi-Key for their help in compiling parts codes.

As always, it's been a pleasure working with the professionals at O'Reilly, in particular Jeff Bleiel, Heather Scherer, and of course, Brian Jepson.

## CHAPTER 1

Theory

1.0 Introduction

Although this book is fundamentally about practice rather than theory, there are a few theoretical aspects of electronics that are almost impossible to avoid.

In particular, if you understand the relationship between voltage, current, and resistance many other things will make a lot more sense.

Similarly, the relationship between power, voltage, and current crops up time and time again.

1.1 Understanding Current

Problem

You want to understand what is meant by *current* in electronics.

Solution

As the word current suggests, the meaning of current in electronics is close to that of the current in a river. You could think of the strength of the current in a pipe as being the amount of water passing a point in the pipe every second. This might be measured in so many gallons per second.

In electronics, current is the amount of charge carried by electrons passing a point in a wire per second (Figure 1-1). The unit of current is the ampere, abbreviated as amp or as unit symbol A.

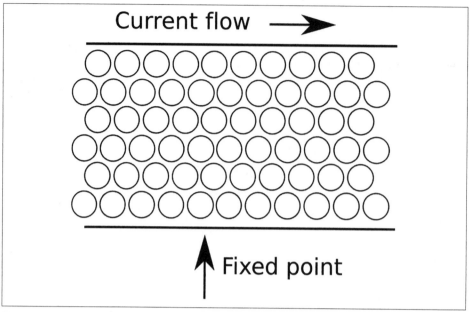

Figure 1-1. Current flowing through a wire

Discussion

For many circuits, one whole amp is quite a large current, so you will see the units of mA (milliamp, a thousandth of an amp) a lot.

See Also

For a list of units and unit prefixes such as mA, see Appendix D.

To learn more about current in a circuit, see Recipe 1.4.

1.2 Understanding Voltage

Problem

You want to understand what is meant by *voltage* in electronics.

Solution

In Recipe 1.1 you read how current is the rate of flow of charge. That current will not flow without something influencing it. In a water pipe, that might be because one end of the pipe is higher than the other.

To understand voltage, it can be useful to think of it as being similar to height in a system of water pipes. Just like height, it is relative, so the height of a pipe above sea level does not determine how fast the water flows through a pipe, but rather how much higher one end of the pipe is than the other (Figure 1-2).

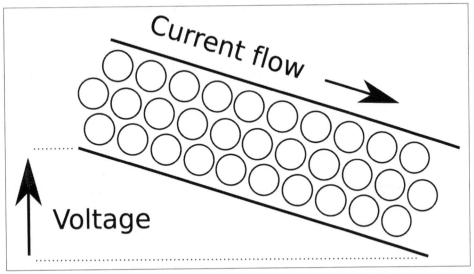

Figure 1-2. Voltage by analogy to height

Voltage might refer to the voltage across a wire (from one end to the other) and in other situations, it might refer to the voltage from one terminal of a battery to another. The common feature is that for voltage to make any sense, it must refer to two points; the higher voltage is the positive voltage, marked with a +.

It is the difference in voltage that makes a current flow in a wire. If there is no difference in voltage between one end of a wire and another then no current will flow.

The unit of voltage is the volt. An AA battery has about 1.5V across its terminals. An Arduino operates at 5V, while a Raspberry Pi operates at 3.3V, although it requires a 5V supply that it reduces to 3.3V.

Discussion

Sometimes it seems like voltage is used to refer to a single point in an electronic circuit rather than a difference between two points. In such cases the voltage then means the difference between the voltage at one point in the circuit and ground. Ground (usually abbreviated as GND) is a local reference voltage against which all other voltages in the circuit are measured. This is, if you like, 0V.

See Also

To learn more about voltages, see Recipe 1.5.

1.3 Calculate Voltage, Current, or Resistance

Problem

You want to understand how the voltage across something controls the current flowing through it.

Solution

Use Ohm's Law.

Ohm's Law states that the current flowing through a wire or electronic component (I) will be the voltage across that wire or component (V) divided by the resistance of the component (R). In other words:

$$I = \frac{V}{R}$$

If it is the voltage that you want to calculate, then this formula can be rearranged as:

$$V = I \times R$$

And, if you know the current flowing through a resistor and the voltage across the resistor, you can calculate the resistance using:

$$R = \frac{V}{I}$$

Discussion

Resistance is the ability of a substance to resist the flow of current. A wire should have low resistance, because you do not usually want the electricity flowing through the wire to be unnecessarily impeded. The thicker the wire, the less its resistance for a given length. So a few feet of thin wire that you might find connecting a battery to a lightbulb (or more likely LED) in a flashlight might have a resistance of perhaps 0.1Ω to 1Ω, whereas the same length of thick AC outlet cable for a kettle may have a resistance of only a couple of milliohms ($m\Omega$).

It is extremely common to want to limit the amount of current flowing through part of a circuit by adding some resistance in the form of a special component called a resistor.

Figure 1-3 shows a resistor (zig-zag line) and indicates the current flowing through it (I) and the voltage across it (V).

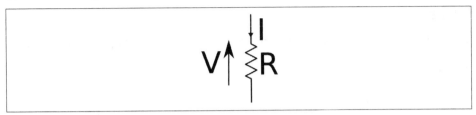

Figure 1-3. Voltage, Current, and Resistance

Let's say that we were to connect a 1.5V battery to a 100Ω resistor as shown in Figure 1-4. The Greek letter Ω (omega) is used as shorthand for the unit of resistance (the "Ohm").

Figure 1-4. Battery and Resistor

Using Ohm's Law, the current is the voltage across the resistor divided by the resistance of the resistor (we can assume that the wires have a resistance of zero).

So, I = 1.5 / 100 = 0.015 A or 15mA.

See Also

To understand what happens to current flowing through resistors and wires in a circuit, see Recipe 1.4.

To understand the relationship between current, voltage, and power, see Recipe 1.6.

1.4 Calculate Current at Any Point in a Circuit

Problem

You want to understand what current will be flowing through any point on a circuit.

Solution

Use Kirchhoff's Current Law.

Stated simply, Kichhoff's Current Law says that at any point on a circuit, the current flowing into that point must equal the current flowing out.

Discussion

For example, in Figure 1-5 two resistors are in parallel and supplied with a voltage from a battery (note the schematic symbol for a battery on the left of Figure 1-5).

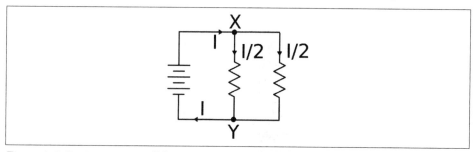

Figure 1-5. Resistors in parallel

At point X, a current of I will be flowing into point X from the battery, but there are two branches out of X. If the resistors are of equal value then each branch will have half the current flowing through it.

At point Y, the two paths recombine and so the two currents of I/2 flowing into Y will be combined to produce a current of I flowing out of Y.

See Also

For Kirchhoff's Voltage Law, see Recipe 1.5.

For further discussion on resistors in parallel, see Recipe 2.5.

1.5 Calculate the Voltages Within Your Circuit

Problem

You want to understand how the voltages around a circuit add up.

Solution

Use Kirchhoff's Voltage Law.

This law states that all the voltages between various points around a circuit will add up to zero.

Discussion

Figure 1-6 shows two resistors in series with a battery. It is assumed that the two resistors are of equal value.

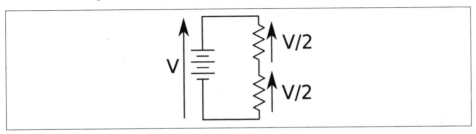

Figure 1-6. Resistors in Series

At first glance it is not clear how Kirchhoff's Voltage Law applies until you look at the polarity of the voltage. On the left, the battery supplies V volts, which is equal in magnitude, but opposite in direction (and hence sign) to the two voltages V/2 across each resistor.

Another way to look at this is that V must be balanced by the two voltages V/2. In other words, $V = V/2 + V/2$ or $V - (V/2 + V/2) = 0$.

See Also

This arrangement of a pair of resistors is also used to scale voltages down (see Recipe 2.6).

For Kirchhoff's Current Law, see Recipe 1.4.

1.6 Understanding Power

Problem

You want to know what is meant by *power* in electronics.

Solution

In electronics, power is the rate of conversion of electrical energy to some other form of energy (usually heat). It is measured in Joules of energy per second, which is also known as a watt (W).

When you wire up a resistor as shown back in Figure 1-4 of Recipe 1.3 the resistor will generate heat and if it's a significant amount of heat then the resistor will get hot. You can calculate the amount of power converted to heat using the formula:

$$P = I \times V$$

In other words, the power in watts is the voltage across the resistor (in volts) multiplied by the current flowing through it in amps. In the example of Figure 1-4 where the voltage across the resistor is 1.5V and the current through it was calculated as 15mA, the heat power generated will be 1.5V x 15mA = 22.5mW.

Discussion

If you know the voltage across the resistor and the resistance of the resistor, then you can combine Ohm's Law and P=IV and use the formula:

$$P = \frac{V^2}{R}$$

With V=1.5V and R of 100Ω the power is 1.5V x 1.5V / 100Ω = 22.5mW.

See Also

For Ohm's Law see Recipe 1.3.

1.7 Alternating Current

Problem

You have heard that electricity comes in two flavors: direct current (DC) and alternating current (AC), and want to know the difference.

Solution

In all the recipes up to this point, DC is assumed. The voltage is constant and generally what you would expect a battery to supply.

AC is what is supplied by wall outlets, and although it can be reduced to lower voltages (see Recipe 3.9) it is generally of a high (and dangerous) voltage. In the US, this means 110V and in most of the rest of the world 220V or 240V.

Discussion

What puts the *alternating* in alternating current is the fact that the direction of current flow in AC reverses many times per second. Figure 1-7 shows how the voltage varies in a US AC wall outlet.

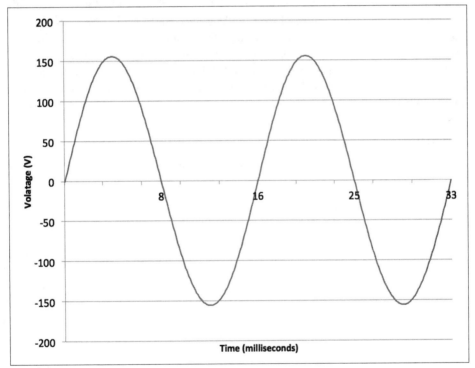

Figure 1-7. Alternating Current

The first thing to notice is that the voltage follows the shape of a sine wave, gently increasing until it exceeds 150V then heading down past 0V to around –150V and then back up again, taking about 16.6 thousandths of a second (milliseconds, or ms) to complete one full cycle.

The relationship between the period of AC (time taken for one complete cycle) and the frequency of the AC (number of cycles per second) is:

$$frequency = \frac{1}{period}$$

The unit of frequency is the Hertz (abbreviated as Hz) so you can see that the AC shown in Figure 1-7 has a period of 16.6ms, which is 0.0166 seconds. So you can calculate the frequency as:

$$frequency = \frac{1}{period} = \frac{1}{0.0166} \approx 60Hz$$

You may be wondering why AC from an outlet is described as 110V when it actually manages to swing over a range of over 300V from peak to peak. The answer is that the 110V figure is the equivalent DC voltage that would be capable of providing the same amount of power. This is called the RMS (root mean square) voltage and is the peak voltage divided by the square root of 2 (which is roughly 1.41). So, in the preceding example, the peak voltage of 155V when divided by 1.41 gives a result of roughly 110V RMS.

See Also

You will find more information on using AC in Chapter 7.

## CHAPTER 2

Resistors

2.0 Introduction

Resistors are used in almost every electronic circuit, come in a huge variety of shapes and sizes, and are available in a range of values that spans milliohms (thousandths of an ohm) to mega ohms (millions of ohms).

Ohm, the unit of resistance, is usually abbreviated as the Greek letter omega (Ω), although you will sometimes see the letter R used instead. For example, 100Ω and 100R both mean a resistor with a resistance of 100 ohms.

2.1 Read Resistor Packages

Problem

You want to work out the value of a resistor.

Solution

On a through-hole resistor (a resistor with leads) that has colored stripes on it, use the resistor color code.

If your resistor has stripes in the same positions as Figure 2-1 then the three stripes together on the left determine the resistor's value and the single stripe on the right determines the accuracy of the value.

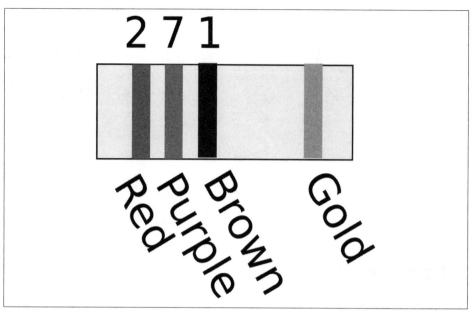

Figure 2-1. A Three-Stripe Resistor

Each color has a value as listed in Table 2-1.

Table 2-1. Resistor Color Codes

Black	0
Brown	1
Red	2
Orange	3
Yellow	4
Green	5
Blue	6
Violet	7
Gray	8
White	9
Gold	1/10
Silver	1/100

For a three-stripe resistor such as this, the first two stripes determine the basic value (say 27 in Figure 2-1) and the third stripe determines the number of zeros to add to the end. In the example of Figure 2-1, the value of a resistor with stripes red, purple, and brown is 270Ω. I said before that this stripe indicates the number of zeros, but actually, to be more accurate, it is a multiplier. If it has a value of gold, then this means $\frac{1}{10}$ of the value indicated by the first two stripes. So brown, black, and gold would indicate a 1Ω resistor.

The stripe on its own specifies the tolerance of the resistor. Silver (rare these days) indicates ±10%, gold ±5%, and brown ±1%.

If your resistor has stripes as shown in Figure 2-2 then the value of the resistor is specified with an extra digit of precision. In this case, the first three stripes determine the basic value (in the case of Figure 2-2, 270) and the final digit the number of zeros to add (in this case, 0). This resistor is also 270Ω.

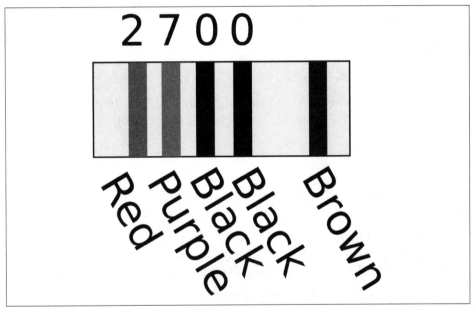

Figure 2-2. A Four-Stripe Resistor

For low-value resistors, gold is used as a multiplier of 0.1 and silver of 0.01. A 1Ω four-stripe resistor would have value stripes of brown, black, black, and silver (100 x 0.01).

Discussion

Although tiny, surface mount technology (SMT) resistors normally have their value of resistance printed on them. However, the same system of base value followed by multiplier is used so a 270Ω SMT resistor would have the number 2700 printed on it and a 1kΩ resistor 1001.

See Also

Through-hole capacitors also have value labels similar to SMT resistors (see Recipe 3.3).

2.2 Find Standard Resistor Values

Problem

You've done your calculations and you need a 239Ω resistor, but how do you match this to a standard value that you can actually buy?

Solution

Buy a resistor from ±5% E24 series.

Values in the E24 series have basic values of 10, 11, 12, 13, 15, 16, 18, 20, 22, 24, 27, 30, 33, 36, 39, 43, 47, 51, 56, 62, 68, 75, 82, and 91, with as many zeros after them as you need.

Discussion

The ±1% E96 series includes all the base values of the E24 series, but has four times as many values. However, it is very rare to need such precise values of resistor.

If your resistor is to limit current to some other component that might be damaged, perhaps limiting the power to an LED (Recipe 4.4) or into the base of a bipolar transistor (Recipe 5.1), then pick the next largest value of resistor from the E24 series.

For example, if your calculations tell you that the resistor should be a 239Ω resistor, then pick a 240Ω resistor from the E24 series.

In reality, you may decide to limit yourself even further to avoid having to gradually collect every conceivable value of resistor, especially as they are often sold in packs of 100. I generally keep the following resistor values in stock: 10Ω, 100Ω, 270Ω, 470Ω, 1k, 3.3k, 4.7k, 10k, 100k, and 1M.

See Also

For full details on all of the resistor series available, see *http://www.logwell.com/tech/components/resistor_values.html*.

2.3 Select a Variable Resistor

Problem

You want to understand how variable resistors work.

Solution

A variable resistor, a.k.a. a *pot* or *potentiometer*, is made from a resistive track and a slider that varies its position along the track. By varying the position of the slider you can vary the resistance between the slider and either of the terminals at the ends of the track. The most common pots are rotary like the one shown in Figure 2-3.

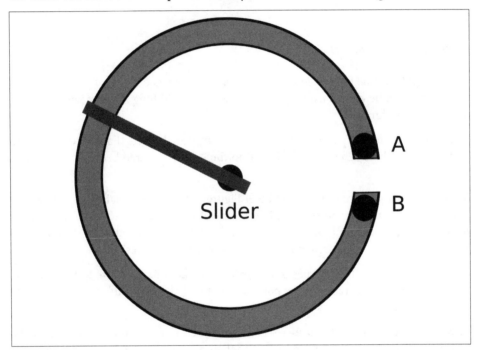

Figure 2-3. A Rotary Pot

Discussion

Pots come in a large variety of shapes and sizes. Figure 2-4 shows a selection of pots.

Figure 2-4. Potentiometers

The two pots on the left of Figure 2-4 are called *trimmers* or *trimpots*. These devices are designed to be turned using a screwdriver or by twiddling the tiny knob between the thumb and forefinger.

The next pot is a pretty standard device with a threaded barrel that allows the pot to be fixed into a hole. The shaft can be cut to the length required before fixing a knob to it.

There is a dual-gang pot in the middle of Figure 2-4. This is actually two pots with a common shaft and is often employed in stereo volume controls. After that is a similar looking device that combines a pot with an on/off switch. Finally, on the right is a sliding type of pot, the kind that you might find on a mixing desk.

Pots are available with two types of tracks. Linear tracks have a close to linear resistance across the whole range of the pot. So, at the half-way point the resistance will be half of the full range.

Pots with a logarithmic track increase in resistance as a function of the log of the slider position rather than in proportion to the position. This makes them more suitable to volume controls as human perception of loudness is logarithmic. Unless you are making a volume control for an audio amplifier, you probably want a linear pot.

See Also

To connect a pot to an Arduino or Raspberry Pi, see Recipe 12.9.

A potentiometer lends itself to being a variable voltage divider (see Recipe 2.6).

2.4 Combine Resistors in Series

Problem

You want to understand the overall resistance and power-handling implications of placing a number of resistors in series.

Solution

The overall resistance of a number of resistors in series is just the sum of the separate resistances.

Discussion

Figure 2-5 shows two resistors in series. The current flows through one resistor and then the second. As a pair, the resistors will be equivalent to a single resistor of 200Ω.

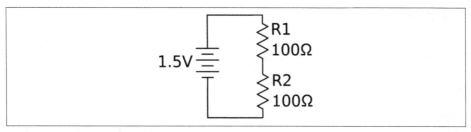

Figure 2-5. Resistors in Series

The heating power of each resistor will be $Power = \frac{V^2}{R} = \frac{0.75^2}{100\Omega} = 5.6mW$.

If you used a single resistor of 200Ω then the power would be:

$$Power = \frac{V^2}{R} = \frac{1.5^2}{200} = 11.3mW$$

So, by using two resistors, you can double the power.

You may be wondering why you would ever want to use two resistors in series when you could just use one. Power dissipation may be one reason, if you cannot find resistors of sufficient power.

But there are other situations, such as the one shown in Figure 2-6, where you use a variable resistor (pot) with a fixed resistor to make sure that the combination's resistance does not fall below the value of the fixed resistor.

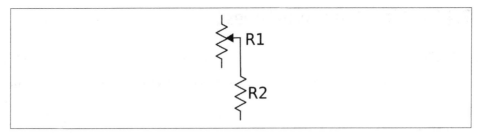

Figure 2-6. A Pot and Fixed Resistor

See Also

Resistors in series are often used to form a voltage divider (see Recipe 2.6).

2.5 Combine Resistors in Parallel

Problem

You want to understand the overall resistance and power-handling implications of placing a number of resistors in parallel.

Solution

The combined resistance of a number of resistors in parallel is the inverse of the sum of the inverses of the resistors. That is, if there are two resistors R1 and R2 in parallel, then the overall resistance is given by:

$$R_{total} = \frac{1}{\frac{1}{R_1} + \frac{1}{R_2}}$$

Discussion

In the example shown in Figure 2-7 with two 100Ω resistors in parallel, the arrangement is equivalent to a single resistor of:

$$R_{total} = \frac{1}{\frac{1}{100\Omega} + \frac{1}{100\Omega}} = \frac{1}{\frac{2}{100\Omega}} = \frac{1}{\frac{1}{50\Omega}} = 50\Omega$$

Intuitively, this makes perfect sense, because there are now two equally resistive paths through the resistors instead of one as would be the case with a single resistor.

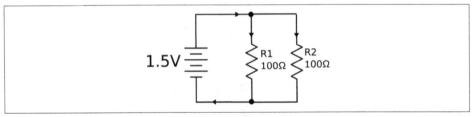

Figure 2-7. Resistors in Parallel

In Figure 2-7 a single 50Ω resistor is equivalent to the two 100Ω resistors in parallel, but what implications does this have for the power ratings of the two resistors?

Intuitively, you would expect the total power dissipation of two 100Ω resistors to be the same as the two 50Ω resistors, but just to be sure let's do the math.

For each 100Ω resistor, the power will be:

$$Power = \frac{V^2}{R} = \frac{1.5^2}{100\Omega} = 22.5mW$$

So the total for the two resistors will be 45mW, allowing you to use lower power and more common resistors.

Just as you'd expect, performing the calculation for the single 50Ω resistor you get:

$$Power = \frac{V^2}{R} = \frac{1.5^2}{50\Omega} = 45mW$$

See Also

See Recipe 2.4 for resistors in series.

2.6 Reduce a Voltage to a Measurable Level

Problem

You have a DC or AC voltage that you want to reduce.

Solution

Use two resistors in series as a voltage divider (also called potential divider). The word "potential" indicates the voltage has the potential to do work and make current flow.

Figure 2-8 shows a pair of resistors being used as a voltage divider.

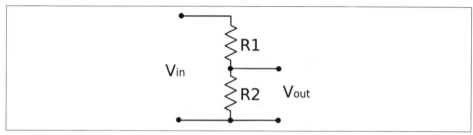

Figure 2-8. A Voltage Divider

The output voltage (Vout) will be a fraction of the input voltage (Vin) according to the formula:

$$V_{out} = \frac{R2}{R1 + R2} \times V_{in}$$

For example, if R1 is 470Ω, R2 is 270Ω, and Vin is 5V:

$$V_{out} = \frac{R2}{R1 + R2} \times V_{in} = \frac{470}{270 + 470} \times 5V = 3.18V$$

Discussion

Note that if R1 and R2 are equal then the voltage is divided by 2.

A pot naturally forms a potential divider as you can think of it as two resistors in series whose overall resistance adds up to the same value, but the proportion of the resistance of R1 to R2 varies as you turn the knob. This is just how a pot is used in a volume control.

It may be tempting to think of a voltage divider as useful for reducing voltages in power supplies. However, this is not the case, because as soon as you try to use the output of the voltage divider to power something (a load), it is as if another resistor is being put in parallel with R2. This effectively reduces the resistance of the bottom half of the voltage divider and therefore also reduces the output voltage. This will only work if R1 and R2 are much lower than the resistance of the load. This makes them great for reducing signal levels, but of no use for high-power circuits.

See Also

See Chapter 7 for various techniques for reducing voltages in power supplies.

For level shifting with a voltage divider, see Recipe 10.17.

2.7 Choose a Resistor that Won't Burn Out

Problem

You want to know what power rating to use for a resistor to avoid overheating and failing it.

Solution

Calculate the power (Recipe 1.6) your resistor will be converting into heat and pick a resistor with a power rating comfortably greater than this value.

For example, if you have a 10Ω resistor connected directly to the terminals of a 1.5V battery, then the power that the resistor converts to heat can be calculated as:

$$P = \frac{V^2}{R} = \frac{1.5 \times 1.5}{10} = 0.225W$$

This means that a standard ¼W resistor will be OK, but you may wish to take the next step up to ½W.

Discussion

The most common power rating of resistor used by hobbyists is the ¼W (250mW) resistor. These resistors are not so tiny that they are hard to handle or the leads too thin to make good contact in breadboard (Recipe 20.1) yet they are capable of handling enough power for most uses, such as limiting current to LEDs (Recipe 14.1) or used as voltage dividers for low current (Recipe 2.6).

Other common power ratings for through-hole resistors with leads are ½W, 1W, 2W, 5W, 10W, and even higher power resistors.

Figure 2-9 shows a selection of resistors of different power ratings.

When it comes to the tiny surface-mount "chip resistors" that are surface mounted onto circuit boards, power ratings start much lower.

See Also

To understand power, see Recipe 1.6.

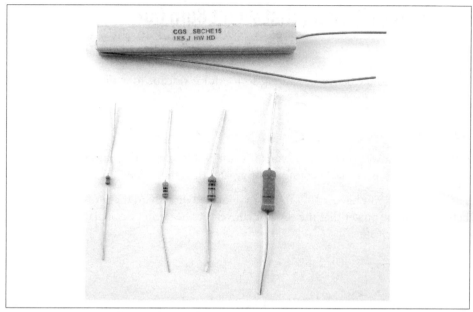

Figure 2-9. Resistors from Left to Right: 0.125W, 0.25W, 0.5W, 1W, and 7W at the Top

2.8 Measure Light Levels

Problem

You want to measure the intensity of light electronically.

Solution

Use a photoresistor.

A photoresistor (Figure 2-10) is a resistor in a clear plastic package whose resistance varies depending on the amount of light falling on it. The more brightly the photoresistor is illuminated, the lower the resistance.

A typical photoresistor might have a resistance in direct sunlight of 1kΩ increasing to several MΩ in total darkness.

Discussion

Photoresistors are often used in a voltage-divider arrangement (Recipe 2.6) with a fixed-value resistor to convert the resistance of the photoresistor into a voltage that can then be used with a microcontroller (Recipe 12.6) or comparator (Recipe 17.10).

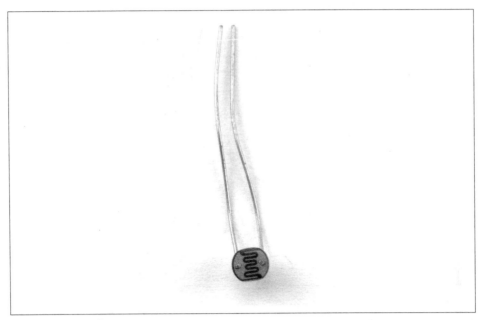

Figure 2-10. A Photoresistor

See Also

You will find more details on using a photoresistor in Recipe 12.6.

2.9 Measure Temperature

Problem

You want to measure temperature electronically.

Solution

One method is to use a thermistor. Other ways can be found in Recipe 12.10 and Recipe 12.11.

All resistors are slightly sensitive to changes in temperature. However, thermistors (Figure 2-11) are resistors whose resistance is very sensitive to changes in temperature. As with photoresistors (Recipe 2.8), they are often used in a voltage divider (Recipe 2.6) to convert the resistance to a voltage reading that is more convenient.

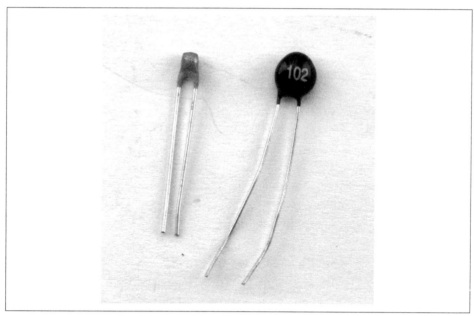

Figure 2-11. Two Thermistors

Discussion

There are two types of thermistors. NTC (negative temperature coefficient) thermistors are the more common type and their resistance decreases as the temperature increases. PTC (positive temperature coefficient) thermistors have a resistance that increases as the temperature increases.

In addition to being used to measure temperature (see Recipe 12.7) PTC thermistors are also used to limit current. As the current through the thermistor increases, the resistor heats up and thus its resistance increases, reducing the current.

See Also

For practical circuits that use a thermistor to measure temperature, see Recipe 12.7 and Recipe 12.8.

2.10 Choose the Right Wires

Problem

The idealized wire has zero resistance. In reality wires do have resistance and you want to know how to account for this in designs as well as know about different types of wire.

Solution

Wires all have resistance. A thick copper wire will have a much lower resistance than the same length of a much thinner wire. As someone once said, "the great thing about standards are that there are always lots to choose from." Nowhere is this more true than for wire thicknesses or gauges. One of the most common standards is AWG (American Wire Gauge) mostly used in the US, and the SWG (Standard Wire Gauge) mostly used in the UK and most logically just the diameter of the wire in mm.

Nearly all wiring used in electronics is made of copper. If you strip the insulation off a wire and it looks silver-colored, then it's probably still copper, but copper that has been "tinned" to stop it from oxidizing and make it easier to solder.

Table 2-2 shows some commonly used wire gauges along with their resistances in Ω per foot and meter for copper wiring.

Table 2-2. Properties of Commonly Used Wire Gauges

AWG	Diameter (mm)	mΩ/m	mΩ/foot	Max. Current (A)	Notes
30	0.255	339	103	0.14	
28	0.376	213	64.9	0.27	
24	0.559	84.2	25.7	0.58	Solid-core hookup wire
19	0.95	26.4	8.05	1.8	General-purpose multicore wire
15	1.8	10.4	3.18	4.7	Thick multicore wire

The larger the AWG number the thicker the wire. Wires thinner than 24AWG are most likely to be thin enameled wire designed for winding transformers and indictors, like the wire shown in Figure 2-12.

Figure 2-12. Enamel-Insulated Wire for Winding Inductors 30-22 AWG

Solid-core wire is made of a single strand of copper insulated in plastic (Figure 2-13). It is useful with solderless breadboards (Recipe 20.1) but should not be used in situations where it is likely to be flexed as it will break through metal fatigue if bent back and forth. I always have this wire in at least three colors so that I can use black for negative, red for positive, and other colors for nonpower connections.

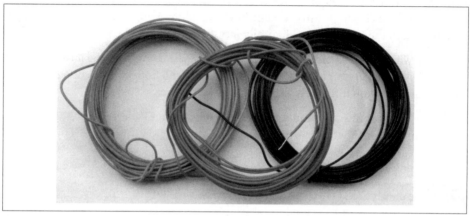

Figure 2-13. Single-core Hookup Wire (24 AWG)

For general-purpose wire where some movement is likely, use multicore wire made up of a number of strands of wire twisted about each other and encased in plastic insulation. Again, it's useful to have a selection of colors.

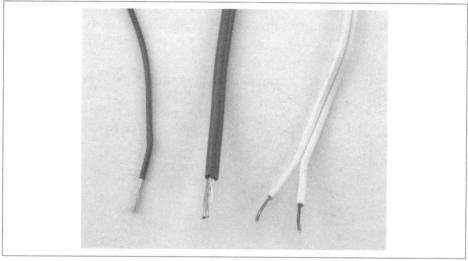

Figure 2-14. 19 and 15 AWG Multicore Wire with Twin "Bell-wire"

Discussion

The currents listed in Table 2-2 are only suggestions. The actual current that each gauge of wire can carry without getting too hot depends on many factors, including how well ventilated the project's case is and whether the wires are all grouped together with a whole load of other wires all getting warm. So treat Table 2-2 as a guide.

When you buy wire, the maximum temperature of the insulation will usually be specified. This is not just because of internal heating of the wire but also for situations where the insulation is to be used in hot environments such as ovens or furnaces.

If you are looking for wires to carry high voltages then you will need a good insulation layer. Again, wires will normally have a breakdown voltage specified for the insulation.

See Also

For a table comparing wire gauges, see *http://bit.ly/2lOyPIh*.

For the current-handling capabilities of wire by gauge, see *http://bit.ly/2mbgZS8*.

## CHAPTER 3

Capacitors and Inductors

3.0 Introduction

When it comes to digital electronics, capacitors are almost a matter of insurance, providing short-term stores of charge that improve the reliability of circuits. As such their use is often simply a matter of following the recommendations in the IC's datasheet without the need to do any math.

However, when it comes to analog electronics the use of capacitors becomes much more varied. Their ability to store small amounts of charge for a short period of time can be used to set the frequency of oscillators (see Recipe 16.5). They can be used to smooth the ripples for a power supply (see Recipe 7.2) or to couple two audio circuits without transferring the DC part of the signal (see Recipe 17.9).

In fact, capacitors will be used throughout this book in all sorts of ways, so it's important to understand how they work, how to select the right capacitor, and how to use it.

Inductors are not as common as capacitors, but they are widely used in certain roles, particularly in power supplies (see Chapter 7).

3.1 Store Energy Temporarily in Your Circuits

Problem

You need an electronic component that can store energy for short periods of time, perhaps to create pulses, or to insulate other components from voltage spikes.

Solution

Use a capacitor.

In construction, capacitors are just two conductive surfaces separated by an insulating layer (Figure 3-1).

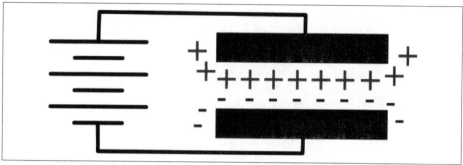

Figure 3-1. A Capacitor

In fact, the insulating layer between the conductive surfaces of the capacitor can just be air, although a capacitor using an air gap will be of very low value. In fact, the value of the capacitor depends on the area of the conducting plates, how close they are together, and how good an insulator separates them. So the greater the area of the plates and the smaller the distance between them the greater the capacitance (the more charge it can store).

Individual electrons do not flow through a capacitor, but those on one side of the capacitor influence those on the other. If you apply a voltage source like a battery to a capacitor, then the plate of the capacitor connected to the positive supply of the battery will accumulate positive charge and the electric field this generates will create a negative charge of equal magnitude on the opposite plate.

In water terms, you can think of a capacitor as an elastic membrane in a pipe (Figure 3-2) that does not allow water to pass all the way through the pipe, but will *stretch*, allowing the capacitor to be charged. If the capacitor is stretched too far then the elastic membrane will rupture. This is why exceeding the maximum voltage of a capacitor is likely to destroy it.

Discussion

When you apply a voltage across a capacitor it will almost instantly charge to that voltage. But if you charge it through a resistor then it will take time to become full of charge. Figure 3-3 shows how a capacitor can be charged or discharged through the use of the switches S1 and S2.

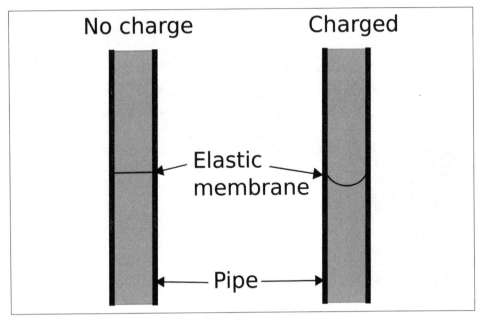

Figure 3-2. Water and Pipe Analogy for a Capacitor

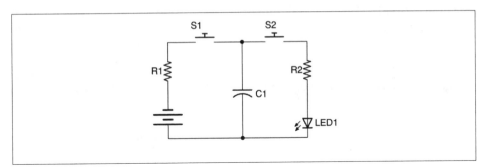

Figure 3-3. Charging and Discharging a Capacitor

When you close switch S1, C1 will charge through R1 until C1 reaches the battery voltage. Open S1 again and the now charged capacitor will retain its charge. Eventually the capacitor will lose its charge through *self-discharge*.

If you now close S2, C1 will now discharge through R2 and LED1, which will light brightly at first and then more dimly as C1 is discharged.

If you want to experiment with the schematic diagram of Figure 3-3, then build up the breadboard diagram shown in Figure 3-4. For an introduction to breadboard, see Recipe 20.1. Use 1kΩ resistors and a 100μF capacitor.

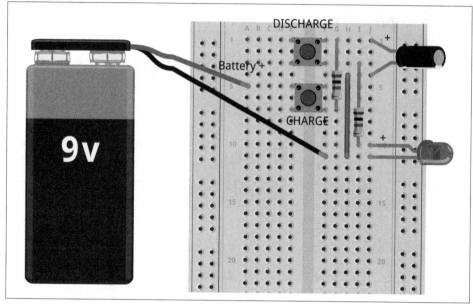

Figure 3-4. Breadboard Layout for Capacitor Experimentation

Press the button labeled CHARGE for a second or two to charge the capacitor then release the button and press the DISCHARGE button. The LED should glow brightly for a second or so and then dim until it extinguishes after a second or so.

If you were to monitor the voltage across the capacitor while it was first being charged and then discharged, you would see something similar to Figure 3-5.

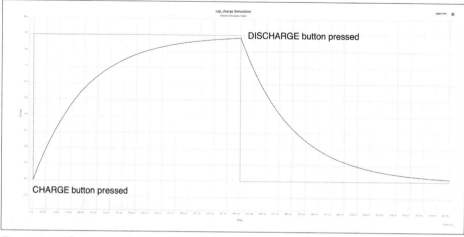

Figure 3-5. Charging and Discharging a Capacitor

In Figure 3-6 the square-shaped waveform is the voltage applied to the capacitor through a 1kΩ resistor. For the first 400ms this is 9V. But as you can see the voltage across the capacitor does not increase linearly, but rather increases faster at first and then gradually tails off as the voltage across the capacitor gets closer to the battery voltage.

Similarly, when the capacitor discharges, the voltage decreases sharply at first and then starts to tail off.

So, if a capacitor stores electrical energy, you may be wondering how this differs from a rechargeable battery. Well, in fact a special type of very high capacitance capacitor called a supercapacitor is sometimes used in place of a rechargeable battery for applications that require a very rapid storage and release of energy. The differences between a capacitor and a battery include:

- A rechargeable battery uses a chemical reaction to generate electricity' a capacitor stores charge directly.
- A rechargeable battery must be charged and discharged over a period of minutes or hours. A capacitor can charge and discharge in fractions of a second.
- The voltage across a capacitor falls off sharply as it starts to discharge, whereas the voltage across a battery stays relatively constant until most of the energy is used.
- Per unit size, a battery can store around ten times as much energy as the best supercapacitor.

See Also

For information on using a solderless breadboard, see Recipe 20.1.

The voltage curves of Figure 3-5 were created using a circuit simulator (Recipe 21.11). You can experiment with this simulation online using PartSim at *http://bit.ly/2mrtrhs*.

3.2 Identify Types of Capacitors

Problem

You need to navigate the bewildering array of capacitor types to select one that is appropriate for your application.

Solution

Unless your application needs capacitors with special features, the following rule of thumb applies.

In most cases, for capacitors between 1pF and 1nF use a disc capacitor (Figure 3-6a). For capacitors between 1nF and 1μF use a multilayer ceramic capacitor (MLC; Figure 3-6b) and for capacitors above 1μF use an aluminum electrolytic capacitor (Figure 3-6c). The rightmost capacitor is a tantalum electrolytic capacitor.

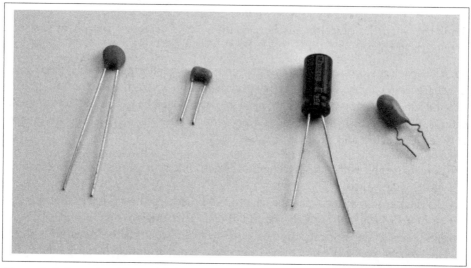

Figure 3-6. Capacitor Types: (a) Disc Ceramic, (b) MLC, (c) Aluminum Electrolytic, and (d) Tantulum

Discussion

Although disc ceramic, MLC, and aluminum electrolytic capacitors are the most commonly used types of capacitor, there are other types:

- Glass and mica capacitors offer a wide temperature range, but are expensive compared to other types of capacitor.
- Tantulum electrolytic capacitors are polarized capacitors that have values around the overlap between the ranges of MLC and electrolytic capacitors. They are small, but relatively expensive and available in values up to a few tens of μF. They suffer from the disadvantage that when they fail, they generally fail in such a way that the terminals of the capacitor become connected to each other, often with fairly explosive consequences. Improvements in MLC capacitors have led to upper values of capacitance of hundreds of μF making tantalum capacitors something of a rarity.

Capacitors are less reliable than resistors. Exceed the voltage rating and you are likely to damage the insulating layer. Electrolytic capacitors use an electrolyte contained in an aluminum can that creates a very thin layer of oxide as the insulator. These are especially prone to failure due to overvoltage, or overtemperature, or just age. If vintage HiFi equipment fails, it is usually the large electrolytic capacitors in the power supplies that are the cause of the failure. In addition, if an electrolytic capacitor does fail then it can spray out the electrolyte in a somewhat messy manner.

Voltage rating

In addition to the actual capacitance of a capacitor, there are a number of other factors to consider when selecting a device. Of critical importance is the voltage rating. Unless you are creating something that uses high voltages, this is not usually a problem with smaller value capacitors, as they are generally rated at least 50V. However, as soon as you get into the range of electrolytics then you will be in a trade-off between capacitor size and voltage. Electrolytic capacitors are commonly available in voltage ratings of 6.3V, 10V, 25V, 30V, 40V, 50V, 63V, 100V, 160V, 200V, 250V, 400V, and 450V. It is unusual to find electrolytic capacitors with a higher voltage rating than 500V.

Temperature rating

Ceramic and MLC capacitors are generally rated for a wide range of temperatures, but aluminum electrolytics are far less tolerant of high temperatures, and are typically rated as 80 or 105° C.

ESR (equivalent series resistance)

The temperature rating becomes important when the capacitors are being rapidly charged and discharged, as a capacitor will always have an internal resistance called its equivalent series resistance (ESR) that causes heating during charging and discharging.

Small-value MLC capacitors generally have a very low ESR of little more than the resistance caused by their leads. This allows them to charge and discharge extremely quickly. A high-value electrolytic capacitor might have an ESR of a few hundred mΩ. This both limits the speed at which the capacitors can charge and discharge and causes a heating effect.

See Also

For the use of electrolytic capacitors to smooth out voltage ripple from power supplies, see Recipe 7.4.

3.3 Read Capacitor Packages

Problem

You have a capacitor and want to identify its value.

Solution

Small, low-value SMD capacitors are generally unmarked, and so you should label them as soon as you buy them.

Electrolytic capacitors usually have their capacitance value and voltage rating printed on the package. Polarized through-hole electrolytic capacitors are also usually supplied with the positive lead longer than the negative lead and the negative lead marked with a minus sign or a diamond symbol.

Most other capacitors use a numbering system similar to that of SMD resistors. The value is usually three digits and a letter. The first two digits are the base value and the third digit is the number of zeros to follow. The base value being the value in pF (pico Farad; see Appendix D).

For example, a 100pF capacitor would have the three digits 101 (a 1, a 0, followed by one further 0). A 100nF capacitor would be marked 104 (a 1, a 0, followed by four further 0s); that is, 100,000 pF or 100nF.

The letter after the digits indicates the tolerance (J, K, or M for ±5%, ±10%, and ±20%, respectively).

Discussion

The number of standard values for capacitors is much smaller than for resistors. Generally the values available are 10, 15, 22, 33, 47, and 68 followed by the necessary number of zeros.

See Also

To read resistor color codes, see Recipe 2.1.

3.4 Connect Capacitors in Parallel

Problem

You want to combine capacitors to increase their overall capacitance.

Solution

Looking at Figure 3-7, you can see two capacitors in parallel double up on the surface area of the conductive plates and therefore may correctly assume that the overall capacitance is the sum of the two capacitances.

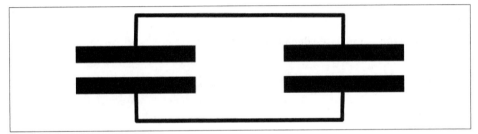

Figure 3-7. Two Capacitors in Parallel

Discussion

It is actually quite common to place a number of capacitors in parallel to increase the overall capacitance. It is especially common when smoothing a high-power transformer-based power supply for, say, an audio amplifier where it is important to remove as much ripple from the power supply as possible (see Recipe 7.2).

In such systems it is common to use a number of different capacitors of different types and values in parallel to minimize the effects of ESR (see Recipe 3.2).

See Also

For capacitors in series, see Recipe 3.5.

3.5 Connect Capacitors in Series

Problem

You want to know why someone has combined capacitors in this unusual fashion.

Solution

When you connect two or more capacitors in series, the total value of capacitance is calculated according to the following formula, which, interestingly, is very similar to resistors in parallel:

$$R_{total} = \cfrac{1}{\cfrac{1}{C_1} + \cfrac{1}{C_2}}$$

Discussion

It is unusual to connect capacitors in series. Occasionally, this will be done as part of a more complex circuit such as in Recipe 7.12.

See Also

For capacitors in parallel, see Recipe 3.4.

3.6 Store Huge Amounts of Energy

Problem

Regular capacitors are not big enough for you.

Solution

Supercapacitors are low-voltage capacitors that have extremely high capacitances. They are primarily used as energy-storage devices in situations that would otherwise use rechargeable batteries.

They can have values up into the hundreds of F (Farads). Note that the top of the capacitance range for an aluminum electrolytic is around 0.22F.

Discussion

Low (comparatively) value supercapacitors of perhaps a few F are sometimes used as alternatives to rechargeable batteries or long-life lithium batteries to power ICs in standby mode to retain memory in static RAM that would otherwise be lost, or to power real-time clock (RTC) chips so a device using an RTC IC will keep the time if it's powered off for a while.

Extremely high-value supercapacitors are available that offer an alternative to rechargeable batteries for larger capacity energy storage.

Supercapacitors with values of 500F or more can be bought for just a few dollars. The maximum voltage for a supercapacitor is 2.7V so, for higher capacity use, the capacitors can be placed in series with special protection circuitry to ensure the 2.7V limit is not exceeded as the bank of capacitors are charged.

Supercapacitors generally look like standard aluminum electrolytic capacitors. At present, their energy storage is still quite a long way from that of rechargeable batteries and because they are capacitors, the voltage decreases much more quickly than when discharging a battery.

See Also

See Recipe 3.7 to calculate the energy stored in super capacitors and normal capacitors.

3.7 Calculate the Energy Stored in a Capacitor

Problem

You have charged up a capacitor to a certain voltage and want to know how much energy the capacitor is storing.

Solution

The energy stored in a capacitor in J is calculated as:

$$E = \frac{CV^2}{2}$$

Discussion

Taking a medium-sized electrolytic of 470µF at 35V the energy stored would be:

$$E = \frac{CV^2}{2} = \frac{0.00047F \times 35V^2}{2} = 0.29J$$

which is not very much energy. Since the energy storage is proportional to the square of the voltage, the results for a capacitor of the same value but at 200V are much more impressive:

$$E = \frac{CV^2}{2} = \frac{0.00047F \times 200V^2}{2} = 9.4J$$

For a 500F 2.7V supercapacitor, the results are even more impressive:

$$E = \frac{CV^2}{2} = \frac{500F \times 2.7V^2}{2} = 1822.5J = 1.822kJ$$

By way of comparison, a single 1.5V AA battery of 2000mAH stores around:

$$2A \times 3600s \times 1.5V = 10.8kJ$$

See Also

For information on rechargeable batteries, see Recipe 8.3.

3.8 Modify and Moderate Current Flow

Problem

You need a component that can filter parts of a signal or smooth out fluctuations.

Solution

An inductor is, at its simplest, just a coil of wire. At DC, it behaves just like any length of wire and will have some resistance, but when AC flows through it, it starts to do something interesting.

A change in current in one direction in an inductor causes a change in voltage in the opposite direction. This effect is more marked the higher the frequency of the AC flowing through the inductor. The net result of this is that the higher the frequency of the AC, the more the inductor *resists* the flow of current. So as not to confuse this effect with ordinary resistance, it is called *reactance* but it still has units of Ω.

The reactance X of an inductor can be calculated by the formula:

$$X = 2\pi f L$$

Where f is the frequency of the AC (cycles per second) and L is the inductance of the inductor whose units are the Henry (H). This resistive effect does not generate heat like a resistor, but rather returns the energy to the circuit.

The inductance of an inductor will depend on the number of turns of wire as well as what the wire is wrapped around. So, very low-value inductors may be just a couple of turns of wire without any core (called air-core inductors). Higher value inductors will generally use an iron or more likely ferrite core. Ferrite is a magnetic ceramic material.

The current-carrying capability of an inductor generally depends on the thickness of the wire used in the coil.

Discussion

Inductors are used in switched mode power supplies (SMPS) where they are pulsed at a high frequency (see Recipe 7.8 and Recipe 7.9). They are also used extensively in radio-frequency electronics, where they are often combined with a capacitor to form a tuned circuit (see Chapter 19).

A type of inductor called a choke is designed to let DC pass while blocking the AC part of a signal. This prevents unwanted radio-frequency noise from infiltrating a circuit. You will often find a USB lead has a lump in it near one end. This is a ferrite choke, which is just a cylinder of ferrite material that encloses the cables and increases the inductance of the wire to a level where it can suppress high-frequency noise.

See Also

For more information on the use of inductors in SMPSs, see Recipe 7.8 and Recipe 7.9.

See Recipe 3.9 for more information on transformers.

3.9 Convert AC Voltages

Problem

You need a component that can convert AC voltages.

Solution

A transformer is essentially two or more inductor coils wrapped onto a single core. Figure 3-8 shows the schematic diagram for a transformer that also gives a clue as to how it works.

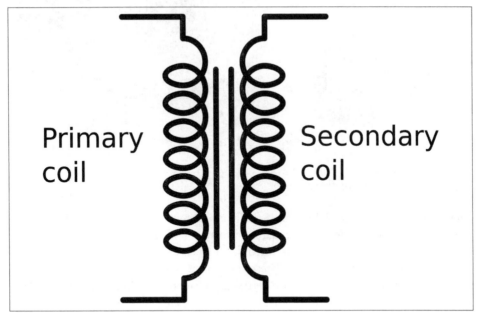

Figure 3-8. The Transformer

A transformer has primary coils and secondary coils. Figure 3-8 shows one of each. The primary coil is driven by AC, say the 110V from an AC outlet, and the secondary is connected to the load.

The voltage at the secondary is determined by the ratio of the number of turns on the primary to the number of turns on the secondary. Thus if the primary has 1000 turns and the secondary just 100, then the AC voltage will be reduced by a factor of 10.

Figure 3-9 shows a selection of transformers. As you can see they come in a wide range of sizes.

In Figure 3-9 there is a small high-frequency transformer on the left that was taken from a disposable flash camera where it was used to step up pulsed DC (almost AC) from a 1.5V battery to the 400V needed by a Xenon flash tube.

The type of transformer shown in the center is commonly used to step-down 110V AC to a low voltage, say 6V or 9V.

Figure 3-9. A Selection of Transformers

The transformer on the right is also designed to step-down AC outlet voltages to lower voltages. It is called a torroidal transformer and the primary and secondary coils are wrapped around a single torroidal former (donut shaped arrangement of iron layers), on top of each other. These transformers are often used in HiFi equipment where the noise found in most SMPSs is considered too high for high-end HiFi amplifiers.

Discussion

It used to be that if you wanted to power a low-voltage DC appliance such as a radio receiver from an AC outlet, then you would first use a transformer to drop the 100V AC at 60Hz to, say, 9V. You would then rectify and smoother the low-voltage AC into DC.

These days, an SMPS (see Recipe 7.8) is normally used as transformers are expensive and heavy items with iron and long lengths of copper-winding wire. However, transformers are still used in SMPS but operate at a very much higher frequency than 60Hz. Operating at a high frequency (often 100s of kHz) allows the transformers to be much smaller and lighter than low-frequency transformers while retaining good efficiency.

See Also

This video shows a torroidal transformer winding machine in action: *https://youtu.be/82PpCzM2CUg*.

Recipe 7.1 describes how to use a transformer to convert one AC voltage to another.

## CHAPTER 4

Diodes

4.0 Introduction

The first diodes to be used in electronics were cat's whisker detectors used in crystal radios. These comprised a crystal of a semiconductor material (often lead sulphide or silicon). The cat's whisker is simply a bare wire that is held in an adjustable bracket that touches the semiconductor crystal. By carefully moving the whisker around, at certain points of contact the arrangement would act as a diode, only allowing current to flow in one direction. This property is needed in a simple radio receiver to detect the radio signal so that it can be heard (see Chapter 19).

Today diodes are much easier to use and come in all sorts of shapes and sizes.

4.1 Block the Flow of Current in One Direction

Problem

You want a component that allows current to flow one way but not the other.

Solution

A diode is a component that only allows current to flow through it in one direction. It's a kind of one-way valve if you want to think of it in terms of water running through pipes, which of course is a simplification. In reality the diode offers very low resistance in one direction and very high resistance in the other. In other words, the one-way valve restricts the flow a tiny bit when open and also leaks slightly when closed. But most of the time, thinking of a diode as a one-way valve for electrical current works just fine.

There are lots of specialized types of diodes, but let's start with the most common and basic diode, the rectifier diode. Figure 4-1 shows such a diode in a circuit with a battery and resistor.

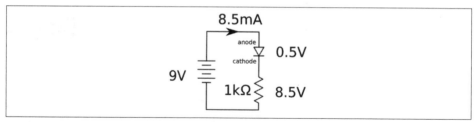

Figure 4-1. A Forward-biased Diode

In this case, the diode allows flow of current and is considered forward-biased. The two leads of the diode are called the anode (abbreviated as "a") and cathode (abbreviated rather confusingly as "k"). For a diode to be forward-biased, the anode needs to be at a higher voltage than the cathode, as shown in Figure 4-1.

One interesting property of a forward-biased diode is that unlike a resistor, the voltage across it does not vary in proportion to the current flowing through it. Instead, the voltage remains almost constant, no matter how much current flows through it. This varies depending on the type of diode, but is generally about 0.5V.

In the case of Figure 4-1, we can calculate that the current flowing through the resistor will be:

$$I = \frac{V}{R} = \frac{9V - 0.5V}{1k\Omega} = 8.5mA$$

This is only 0.5mA less than if the diode were to be replaced with wire.

In Figure 4-2 the diode is facing the other direction. It is reverse-biased and consequently almost no current will flow through the resistor.

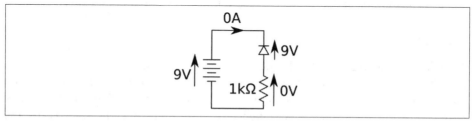

Figure 4-2. A Reverse-biased Diode

Discussion

The diode's one-way effect can be used to convert AC (Recipe 1.7) into DC. Figure 4-3 shows the effect of a diode on an AC voltage source.

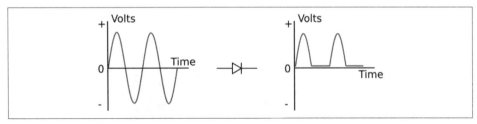

Figure 4-3. Rectification

This effect is called rectification (see Recipe 7.2). The negative part of the cycle is unused. This still isn't true DC because even though the voltage never becomes negative it still swings from 0V up to a maximum and back down rather than remain constant. The next stage would be to add a capacitor in parallel with the load resistor that would smooth out the signal to a flat and almost constant DC voltage.

See Also

For information on using diodes in power supplies, see Recipe 7.2 and Recipe 7.3.

4.2 Know Your Diodes

Problem

You want to know about the different types of diodes and their uses.

Solution

Figure 4-4 shows different types of diodes. Generally, the bigger the diode package, the greater its power-handling capability. Most diodes are a black plastic cylinder with a line at one end that points to the cathode (the end that should be more negative for forward-biased operation).

The diode on the left of Figure 4-4 is an SMD. The through-hole diodes to the right get larger, the higher the current rating.

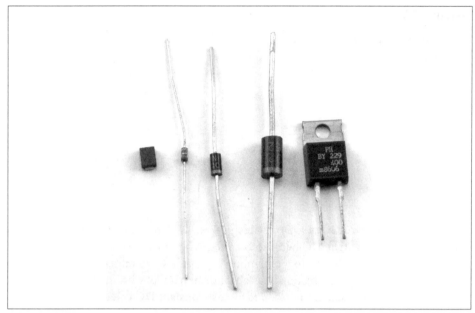

Figure 4-4. A Selection of Diodes

Discussion

There are many different types of diodes. Unlike resistors that are bought as a particular value resistor (say, 1kΩ) diodes are identified by the manufacturer's part number.

Some of the most commonly used rectifier diodes are listed in Table 4-1.

Table 4-1. Common Diodes

Part Number	Typical Forward Voltage	Maximum Current	DC Blocking Voltage	Recovery Time
1N4001	0.6V	1A	50V	30µs
1N4004	0.6V	1A	400V	30µs
1N4148	0.6V	200mA	100V	4ns
1N5819	0.3V	1A	40V	10ns

The forward voltage, often abbreviated as Vf, is the voltage across the diode when forward-biased. The DC-blocking voltage is the reverse-biased voltage that if exceeded may destroy the diode.

The recovery time of a diode refers to how quickly the diode can switch from forward-biased conducting to reverse-biased blocking. This is not instantaneous in any diode and in some applications fast switching is needed.

The 1N5819 diode is called a Schottky diode. These types of diodes have much lower forward voltage and generate less heat.

See Also

You can find the datasheet for the 1N4000 family of diodes here: *http://bit.ly/2lOtD71*.

4.3 Use a Diode to Restrict DC Voltages

Problem

You need to use a diode to allow voltages up to a certain voltage to pass through.

Solution

Use a Zener diode.

When forward-biased, Zener diodes behave just like regular diodes and conduct. At low voltages, when reverse-biased they have high resistance just like normal diodes. However, when the reverse-biased voltage exceeds a certain level (called the breakdown voltage), the diodes suddenly conduct as if they were forward-biased.

In fact, regular diodes do the same thing as Zener diodes but at a high voltage and not a voltage that is carefully controlled. The difference with a Zener diode is that the diode is deliberately designed for this breakdown to occur at a certain voltage (say, 5V) and for the Zener diode to be undamaged by such a "breakdown."

Discussion

Zener diodes are useful for providing a reference voltage (see the schematic in Figure 4-5). Note the slightly different component symbol for a Zener diode with the little arms on the cathode.

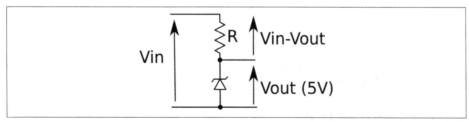

Figure 4-5. Using a Zener Diode to Provide a Reference Voltage

The resistor R limits the current flowing through the Zener diode. This current is always assumed to be much greater than the current flowing into a load across the diode.

This circuit is only well suited to providing a *voltage reference*. A voltage reference provides a stable voltage but with hardly any load current; for example, when the circuit is used with a transistor as in Recipe 7.4. So a resistor value of, say, 1kΩ would, if Vin was 12V, allow a current of:

$$I = \frac{V}{R} = \frac{12-5}{1000} = 7mA$$

The output voltage will remain roughly 5V whatever Vin is as long as it's greater than 5V. To understand how this happens, imagine that the voltage across the Zener diode is less than its 5V breakdown voltage. The Zener's resistance will therefore be high and so the voltage across it due to the voltage divider effect of R and the Zener will be higher than the breakdown voltage. But wait, since the breakdown voltage is exceeded it will conduct, bringing the Vout down to 5V. If it falls below that, the diode will turn off and the Vout will increase again.

Zener diodes are also used to protect sensitive electronics from high-voltage spikes due to static discharge or incorrectly connected equipment. Figure 4-6 shows how an input to an amplifier not expected to exceed ±10V can be protected from both high positive and negative voltages. When the input voltage is within the allowed range the Zener will be high resistance and not interfere with the input signal, but as soon as the voltage is exceeded in either direction the Zener will conduct the excess voltage to ground.

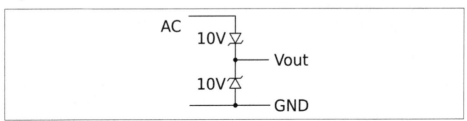

Figure 4-6. Protecting Inputs from Over-Voltage

See Also

Although you would generally use a voltage-regulator IC (see Recipe 7.4), you can use a Zener diode combined with a transistor to act as a voltage regulator.

4.4 Let There Be Light

Problem

You need a component that can generate light without using a lot of power.

Solution

LEDs are like regular diodes in that when reverse-biased they block the flow of current, but when forward-biased they emit light.

The forward voltage of an LED is more than the usual 0.5V of a rectifier and depends on the color of the LED. Generally a standard red LED will have a forward voltage of about 1.6V.

Discussion

Figure 4-7 shows an LED in series with a resistor. The resistor is necessary to prevent too much current from flowing through the LED and damaging it.

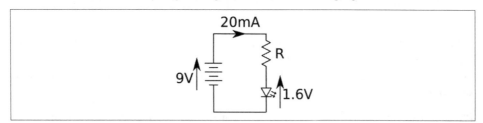

Figure 4-7. Powering an LED

An LED used as an illuminator will generally emit some light at 1mA but usually needs about 20mA for optimum brightness. The LED's datasheet will tell you its optimal and maximum forward currents.

As an example, if in Figure 6-5 the voltage source is a 9V battery and the LED has a forward voltage of 1.6V, you can calculate the resistor value needed using Ohm's Law:

$$R = \frac{V}{I} = \frac{9V - 1.6V}{20mA} = 370\Omega$$

370Ω is not a common resistor value (see Recipe 2.2) so you could pick a 360Ω resistor, in which case the current would be:

$$I = \frac{V}{R} = \frac{9V - 1.6V}{360\Omega} = 20.6mA$$

which would be just fine.

Finding suitable resistor values for LEDs to limit the current is such a common task that there is really no need to go through these calculations all the time. In Recipe 14.1 you will find a practical recipe for rule-of-thumb selection of current-limiting resistors.

See Also

For information on driving different types of LED, see Chapter 14.

4.5 Detect Light

Problem

You want to take a reading of the light level.

Solution

Use a photodiode. You can also use a photoresistor (Recipe 2.8) or a phototransistor (Recipe 5.7).

A photodiode is a diode that is sensitive to light. A photodiode usually has a transparent window, but photodiodes designed for infrared use have a black plastic case. The black plastic case is transparent to IR and usefully stops the photodiode from being sensitive to visible light.

Photodiodes can be treated as little tiny photovoltaic solar cells. When illuminated they generate a small current. Figure 4-8 shows how you can use a photodiode with a resistor to produce a small voltage that you could then use in your circuits.

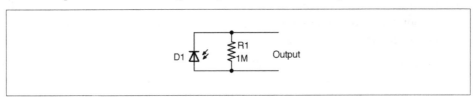

Figure 4-8. A Photodiode in Photovoltaic Mode

In this circuit, the voltage when illuminated brightly might only be 100mV.

The resistor is necessary so that the small current from the photodiode is converted into a voltage (V=IR). Otherwise, any voltage that you measure will be depend on the resistance (called impedance when not from a resistor) of whatever is measuring the voltage. So, for example, a multimeter with an input impedance of 10MΩ would provide a completely different (and lower) reading than a multimeter with a 100MΩ input impedance.

R1 makes the voltage consistent. The impedance of whatever is connected to the output needs to be much higher in value than R1. If this is an op-amp (see Chapter 17) then the input impedance is likely to be hundreds of MΩ and so will not alter the output voltage appreciably. The smaller you make R1, the lower the output voltage will be, so it's a matter of striking a balance.

Better sensitivity can be achieved by using the photodiode in a photoconductive mode with a voltage source (Figure 4-9).

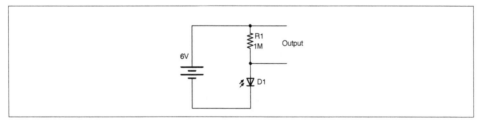

Figure 4-9. A Photodiode in Photoconductive Mode

Discussion

Photodiodes are quite linear, so they are often used in light meters. They also respond pretty quickly and are used in telecommunications systems to sense optical signals.

See Also

Photoresistors (Recipe 12.6) and phototransistors (Recipe 5.7) tend to be used more than photodiodes as they are more sensitive.

## CHAPTER 5

Transistors and Integrated Circuits

5.0 Introduction

Transistors are used to control the flow of a current. In digital electronics this control takes the form of an on/off action, with the transistor acting as an electronic switch.

Transistors are also used in analog electronics where they can be used to amplify signals in a linear manner. However, these days, a better (cheaper and more reliable) way to do this is to use an integrated circuit (chip) that combines lots of transistors and other components into a single convenient package.

This chapter does not cover all types of transistors or semiconductor devices, but instead focuses on the most common ones, which are generally low in cost and easy to use. There are other exotic devices like the unijunction transistors and SCRs (silicon-controlled rectifiers) that used to be popular, but are now seldom used.

The other thing deliberately left out of this cookbook is the usual theory of how semiconductors like diodes and transistors work. If you are interested in the physics of electronics, there are many books and useful resources on semiconductor theory, but just to make use of transistors, you do not really need to know about holes, electrons, and doping N and P regions.

This chapter will concentrate on the use of transistors in their digital role. You will find information on using transistors for analog circuits in Chapter 16.

In this chapter, you will encounter a wide variety of transistors. Appendix A includes pinouts for the transistors used in this chapter and throughout the book.

5.1 Switch a Stronger Current Using a Weaker One

Problem

You want to incorporate digital switches into your project to control whether current flows or not.

Solution

Use a low-cost bipolar junction transistor (BJT).

BJTs like the 2N3904 cost just a few cents and are often used with a microcontroller output pin from an Arduino or Raspberry Pi to increase the current the pin can control.

Figure 5-1 shows the schematic symbol for a BJT alongside one of the most popular models of this kind of transistor, the 2N3904. The 2N3904 is in a black plastic package called a TO-92 and you will find that many different low-power transistor models come supplied in the TO-92 package.

The transistor is usually shown with a circle around it, but sometimes just the symbol inside the circle is used.

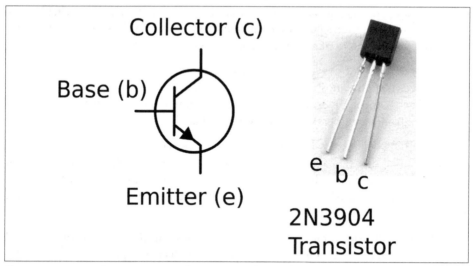

Figure 5-1. A Bipolar Transistor Schematic Symbol and Actual Device

The three connections to the transistor in Figure 5-1 are from top to bottom:

- The collector—the main current to be controlled flows into the collector
- The base—the control connection

- The emitter—the main current flows out through the emitter

The main current flowing into the collector and out of the emitter is controlled by a much smaller current flowing into the base and out of the emitter. The ratio of the base current to the collector current is called the current gain of the transistor and is typically somewhere between 100 and 400. So for a transistor with a *gain* of 100, a 1mA current flowing from base to emitter will allow a current of up to 100mA to flow from collector to emitter.

Discussion

To get a feel for how such a transistor could be used as a switch, build the circuit shown in Figure 5-2 using the breadboard layout shown in Figure 5-3. For help getting started with breadboard tutorial, see Recipe 20.1.

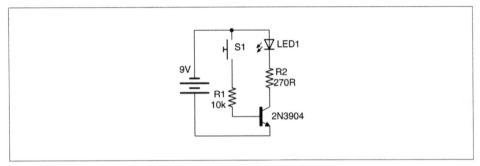

Figure 5-2. A Schematic for Experimenting with a Transistor

The push switch will turn the LED on when its pressed. Although this could be achieved much more easily by just putting the switch in series with the LED and R2, the important point is that the switch is supplying current to the transistor through R1. A quick calculation shows that the maximum current that could possibly flow through R1 and the base is:

$$I = \frac{V}{R} = \frac{9V}{10k} = 0.9mA$$

In reality the current is less than this, because we are ignoring the 0.5V between the base and emitter of the transistor. If you want to be more precise, then the current is actually:

$$I = \frac{V}{R} = \frac{(9V - 0.6V)}{10k} = 0.84mA$$

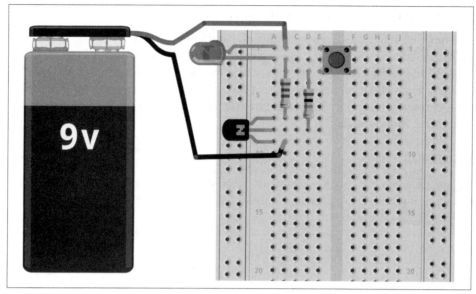

Figure 5-3. A Breadboard Layout for Experimenting with a Transistor

This tiny current flowing through the base is controlling a much bigger current of roughly (assuming Vf of LED 1.8V):

$$I = \frac{V}{R} = \frac{9V - 1.8V}{270} = 26.67mA$$

Just like a diode, when the BJT is in use there will be an almost constant voltage drop of around 0.5V to 1V between the base and emitter connections of the transistor.

Limiting Base Current

A resistor limiting the current that can flow into the base of the transistor (R1 in Figure 5-2) is essential, because if too much current flows through the base then the transistor will overheat and eventually die in a puff of smoke.

Because the base of a transistor only requires a small current to control a much bigger current, it is tempting to think of the base as having a built-in resistance to a large current flowing. This is not the case, since it will draw a self-destructively large current as soon as its base voltage exceeds 0.6V or so. So, always use a base resistor like R1.

The BJT transistor described earlier is the most common and is an NPN type of transistor (negative positive negative). This is not a case of the transistor being indecisive, but relates to the fact that the transistor is made up like a sandwich with N-type (negative) semiconductor as the bread and P-type (positive) semiconductor as the filling. If you want to know what this means and understand the physics of semiconductors, then take a look at *https://en.wikipedia.org/wiki/Bipolar_junction_transistor*.

There is another less commonly used type of BJT called a PNP (positive negative positive) transistor. The filling in this semiconductor sandwich is negatively doped. This means that everything is flipped around. In Figure 5-2 the load (LED and resistor) is connected to the positive end of the power supply and switched on the negative side to ground. If you were to use a PNP transistor the circuit would look like Figure 5-4. You can find out more about using PNP transistors in Recipe 11.2.

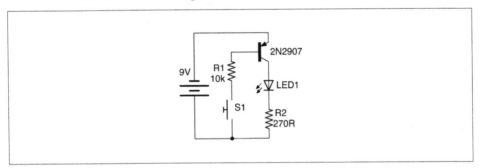

Figure 5-4. Using a PNP Transistor

See Also

If the gain of your transistor is not enough, you may need to consider a Darlington transistor (Recipe 5.2) or MOSFET (Recipe 5.3).

If, on the other hand, you need to switch a high-power load then you should consider a power MOSFET (Recipe 5.3) or IGBT (Recipe 5.4).

You can find the datasheet for the popular low-power BJT, the 2N3904, here: *http://www.farnell.com/datasheets/1686115.pdf*.

5.2 Switch a Current with Minimal Control Current

Problem

You need more gain than you can get with a BJT to switch a current with a very small control current.

Solution

Use a Darlington transistor.

A regular BJT will typically only have a gain (ratio of base current to collector current) of perhaps 100. A lot of the time, this will be sufficient, but sometimes more gain is required. A convenient way of achieving this is to use a Darlington transistor, which will typically have a gain of 10,000 or more.

A Darlington transistor is actually made up of two regular BJTs in one package as shown in Figure 5-5. Two common Darlington transistors are shown next to the schematic symbol. The smaller one is the MPSA14 and the larger the TIP120. See Recipe 5.5 for more information on these transistors.

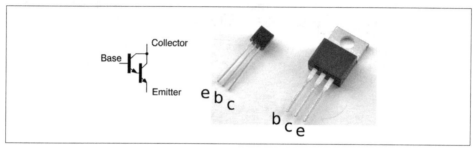

Figure 5-5. Darlington Transistors

The overall current gain of the pair of transistors in this arrangement is the gain of the first transistor multiplied by the gain of the second. It's easy to see why this is the case, as the base of the second transistor is supplied with current from the collector of the first.

Discussion

Although you can use a Darlington transistor in designs just as if it were a single BJT, one effect of arranging the transistors like this is that there are now two voltage drops between the base and emitter. The Darlington transistor behaves like a regular NPN BJT with an exceptionally high gain but a base-emitter voltage drop of twice that of a normal BJT.

A popular and useful Darlington transistor is the TIP120. This is a high-power device that can handle a collector current of up to 5A.

See Also

You can find the datasheet for the TIP120 here: *http://bit.ly/2mHBQy6* and the MPSA14 here: *http://bit.ly/2mI1vXF*.

5.3 Switch High Current Loads Efficiently

Problem

You need to switch really heavy and noisy loads like those found in motors and you need to do it efficiently without generating a lot of heat.

Solution

Use a MOSFET.

MOSFETs (metal-oxide semiconductor field effect transistors) do not have an emitter, base, and collector, but rather a source, gate, and drain. Like BJTs, MOSFETs come in two flavors: N-channel and P-channel. It is the N-channel that is most used and will be described in this recipe. Figure 5-6 shows the schematic symbol for an N-channel MOSFET with a couple of common MOSFETs next to it. The larger transistor (in a package called TO-220) is an FQP30N06 transistor capable of switching 30A at 60V. The hole in the TO-220 package is used to bolt it to a heatsink, something that is only necessary when switching high currents. The small transistor on the right is the 2N7000, which is good for 500mA at 60V.

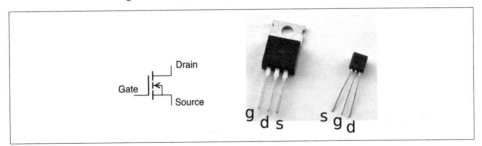

Figure 5-6. MOSFETs

Rather than multiply a current in the way that a BJT does, there is no electrical connection between the gate and other connections of the MOSFET. The gate is separated from the other connections by an insulating layer. If the gate-drain voltage exceeds the threshold voltage of the MOSFET, then the MOSFET conducts and a large current can flow between the drain and source connections of the MOSFET. The threshold voltage varies between 2V and 10V. MOSFETs designed to work with digital outputs from a microcontroller such as an Arduino or Raspberry Pi are called logic-level MOSFETs and have a gate-threshold voltage guaranteed to be below 3V.

If you look at the datasheet for a MOSFET you will see that it specifies on and off resistances for the transistor. An on resistance might be as low as a few mΩ and the off resistance many MΩ. This means that MOSFETs can switch much higher currents than BJTs before they start to get hot.

You can calculate the heat power generated by the MOSFET using the current flowing through it and its on resistance using the following formula. For more information on power, see Recipe 1.6.

$$P = I^2 R_{on}$$

Discussion

The schematic and breadboard layout used in Recipe 5.1 needs a little modification to be used to experiment with a MOSFET. The variable resistor is now used to set the gate voltage from 0V to the battery voltage. The revised schematic and breadboard layouts are shown in Figures 5-7 and 5-8, respectively.

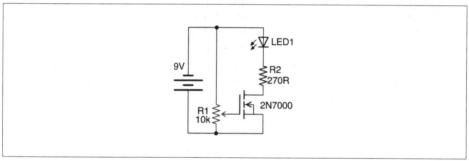

Figure 5-7. Schematic for Experimenting with a MOSFET

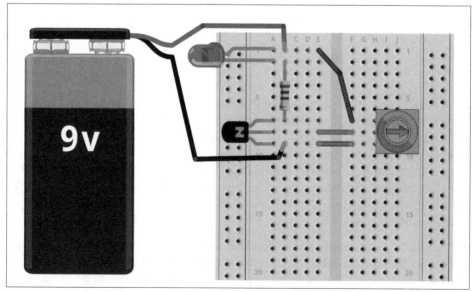

Figure 5-8. Breadboard Layout for Experimenting with a MOSFET

With the trimpot's knob at the 0V end of its track, the LED will not be lit. As the gate voltage increases to about 2V the LED will start to light and when the gate voltage gets to about 2.5V the LED should be fully on.

Try disconnecting the end of the lead going to the slider of the pot and touch it to the positive supply from the battery. This should turn the LED on, and it should stay on even after you take the lead from the gate away from the positive supply. This is because there is sufficient charge sitting on the gate of the MOSFET to keep its gate voltage above the threshold. As soon as you touch the gate to ground, the charge will be conducted away to ground and the LED will extinguish.

Since MOSFETs are voltage- rather than current-controlled devices you might be surprised to find that under some circumstances you do have to consider the current flowing into the gate. That is because the gate acts like one terminal of a capacitor. This capacitor has to charge and discharge and so when pulsed at high frequency the gate current can become significant. Using a current-limiting resistor to the gate will prevent too much current from flowing.

Another difference between using a MOSFET and a BJT is that if the gate connection is left floating, then the MOSFET can turn on when you aren't expecting it. This can be prevented by connecting a resistor between the gate and source of the MOSFET.

See Also

To use MOSFETs with a microcontroller output, see Recipe 11.3.

To use a MOSFET for polarity protection, see Recipe 7.17.

For level shifting using a MOSFET, see Recipe 10.17.

For the use of heatsinks, see Recipe 20.7.

5.4 Switch Very High Voltages

Problem

You need much more power.

Solution

An IGBT (insulated gate bipolar transistor) is an exotic type of transistor found in high-power, high-voltage switching applications. They are fast switching, and generally are particularly well specified when it come to operating voltage. Switching voltages of 1000V are not uncommon.

A BJT has base, emitter, and collector; a MOSFET has a gate, source, and drain; and an IGBT combines the two, having a gate, emitter, and collector. Figure 5-9 shows the schematic symbol for an IGBT alongside two IGBTs. The smaller of the two (STGF3NC120HD) is capable of switching 7A at 1.2kV and the even bigger one (IRG4PC30UPBF) 23A at 600V.

Figure 5-9. The Schematic Symbol for an IGBT and Two IGBTs

Discussion

IGBTs are voltage-controlled devices just like a MOSFET, but the switching side of the transistor behaves just like a BJT. The gate of an IGBT will have a threshold voltage just like a MOSFET.

IGBTs are sometimes used in the same role as high-power MOSFETs but have the advantage over MOSFETs of being able to switch higher voltages at equally large currents.

See Also

For information on BJTs see Recipe 5.1 and for MOSFETS see Recipe 5.3.

You can find the datasheet for the STGF3NC120HD here: *http://bit.ly/2msNM6v*.

The IRG4PC30UPBF datasheet is here: *http://bit.ly/2msXTb9*.

5.5 Choosing the Right Transistor

Problem

There are just so many transistors to choose from, how do you pick the right one?

Solution

Stick to a basic set of go-to transistors for most applications until you have more unusual requirements and then find something that fits the bill.

A good set of go-to transistors is shown in Table 5-1.

Table 5-1. A Useful Set of Transistors

Transistor	Type	Package	Max. Current	Max. Volts
2N3904	Bipolar	TO-92	200mA	40V
2N7000	MOSFET	TO-92	200mA	60V
MPSA14	Darlington	TO-92	500mA	30V
TIP120	Darlington	TO-220	5A	60V
FQP30N06L	MOSFET	TO-220	30A	60V

When shopping for an FQP30N06L, make sure the MOSFET is the "L" (for logic) version, with "L" on the end of the part name; otherwise, the gate-threshold voltage requirement may be too high to connect the gate of the transistor to a microcontroller output especially if it is operating at 3.3V.

The MPSA14 is actually a pretty universally useful device for currents up to 1A, although at that current there is a voltage drop of nearly 3V and the device gets up to a temperature of 120°C! At 500mA, the voltage drop is a more reasonable 1.8V and the temperature 60°C.

To summarize, if you only need to switch around 100mA then a 2N3904 will work just fine. If you need up to 500mA, use an MPSA14. Above that, the FQP30N06L is probably the best choice, unless price is a consideration, because the TIP120 is considerably cheaper.

Transistor Datasheets

When you are designing a circuit, you really need to know exactly how your component will behave before you start experimenting.

All components have datasheets that at the very least describe their absolute maximum ratings. This information tells you what you need to do to effectively destroy the components.

For example, if you look at the datasheet for the 2N3904 transistor, you will see a section called absolute maximum ratings that gives you the information shown in Table 5-2.

Table 5-2. 2N3904 Absolute Maximum Ratings

Symbol	Parameter	Value	Units
Vcbo	Collector-Base Voltage	60	V
Ic	Collector Current	200	mA
Ptot	Total Dissipation at 25 degrees C	625	mW
Tj	Max. Junction Temperature	150	C

As long as you stay below these limits, the datasheet is effectively guaranteeing that the device will work okay. Of course, it's always a good idea to err on the side of caution.

The datasheet will also have a section called electrical characteristics that tells you how the device will work under normal circumstances. For a transistor, the one you will be most interested in is the DC gain. The 2N3904 datasheet DC gain is shown in Table 5-3.

Table 5-3. 2N3904 DC Current Gain

Symbol	Parameter	Test Conditions	Min	Typ	Max	Unit
hFE	DC Current Gain	Ic=0.1mA (all Vce=1V)	60		300	
		Ic=1mA	80			
		Ic=10mA	100			
		Ic=100mA	30			

This tells you that at 10mA the minimum gain you can expect is 100, i.e., with a collector current of 10mA you will only need 0.1mA of base current. Notice how this falls right down to 30 with 100mA collector current. This implies that you will need a base current of 3.33mA.

In reality the gain may be much greater than this, but you should not assume anything better in your designs or you may be relying on the characteristics of one particular transistor. This means design may not work when someone else comes to make it using the same model of transistor from a different manufacturing batch.

Discussion

Reasons for straying from the transistors listed in Table 5-1 include:

- You need to switch at high frequency—look for BJTs or FETs described as RF.
- You need to switch higher voltages—BJTs and MOSFETs are readily available at quite high voltages up to 400V, but for even higher voltages look at IGBTs.
- For high currents, MOSFETs take a beating. Look for devices with the lowest possible resistance as this will be the main thing that determines how hot it gets and therefore how much current it can cope with before it fails.

See Also

For information on BJTs see Recipe 5.1, for MOSFETs Recipe 5.3, and for IGBTs Recipe 5.4.

To decide on a transistor for switching with an Arduino or Raspberry Pi, see Recipe 11.5.

Once you start to push transistors beyond a light load, you will find that they start to get hot. If they get too hot they will burn out and be permanently destroyed. To avoid this, either use a transistor with greater current-handling capability or fix a heatsink to it (Recipe 20.7).

5.6 Switching Alternating Current

Problem

You need a transistor that can switch AC.

Solution

TRIAC (TRIode for alternating current) is a semiconductor-switching device designed specifically to switch AC.

BJT and MOSFETs are not useful for switching AC. They can only do that if you split the positive and negative halves of the cycle and switch each with a separate transistor. It is far better to use a TRIAC that can be thought of as a switchable pair of back-to-back diodes.

Figure 5-10 shows how a TRIAC might be used to switch a high AC load using a small current switch. A circuit like this is often used so that a small low-power mechanical switch can be used to switch a large AC current.

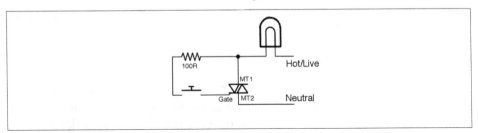

Figure 5-10. Switching AC with a TRIAC

Switching High-Voltage AC

The 110V or 220V AC that you find in your home kills thousands of people every year. It can burn you and it can stop your heart, so please do not attempt to build the circuits described here unless you are absolutely sure that you can do so safely.

Also, see Recipe 21.12.

Discussion

When the switch is pressed a small current (tens of milliamps) flows into the gate of the TRIAC. The TRIAC conducts and will continue conducting until the AC crosses zero volts again. This has the benefit that the power is switched off at low power when the voltage is close to zero, reducing the power surges and electrical noise that would otherwise occur if switching inductive loads like motors.

However, the load could still be switched on at any point in the circuit, generating considerable noise. Zero-crossing circuits (see Recipe 5.9) are used to wait until the next zero crossing of the AC before turning the load on, reducing electrical noise further.

See Also

See Recipe 1.7 on AC.

For information on solid-state relays using TRIACs, see Recipe 11.10.

TRIACs are usually supplied in TO-220 packages. For pinouts, see Appendix A.

5.7 Detecting Light with Transistors

Problem

You want to measure light levels using something other than a photoresistor or photodiode.

Solution

A phototransistor is essentially a regular BJT with a translucent top surface that allows light to reach the actual silicon of the transistor. Figure 5-11 shows how you can use a phototransistor to produce an output voltage that varies as the level of light falling on the transistor changes.

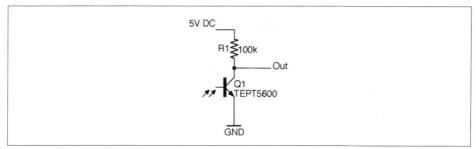

Figure 5-11. Using a Phototransistor

When the photoresistor is brightly illuminated, it will turn on and conduct, pulling the output voltage toward 0V. In complete darkness, the transistor will be completely off and the output voltage will rise to the supply voltage of 5V.

Discussion

Some phototransistors look like regular transistors with three pins and a clear top and others (like the TEPT5600 used in Figure 5-11) come in LED-like packages with pins for the emitter and collector, but no pin for the base. For phototransistors that look like LEDs, the longer pin is usually the emitter. Phototransistors are nearly always NPN devices.

As a light-sensing element, the phototransistor is more sensitive than a photodiode and responds more quickly than a photoresistor. Phototransistors have the advantage over photoresistors that they are manufactured using cadmium sulphide and some countries have trade restrictions on devices containing this material.

See Also

The output of the circuit of Figure 5-11 can be directly connected to an Arduino's analog input (Recipe 10.12) to take light readings as an alternative to using a photoresistor (Recipe 12.3 and Recipe 12.6).

For the TEPT5600 datasheet, see *http://bit.ly/2m8vhS0*.

5.8 Isolating Signals for Safety or Noise Elimination

Problem

For safety or noise immunity reasons, you want a signal to flow from one part of a circuit to another without there being any electrical connection.

Solution

Use an opto-coupler, which consists of an LED and phototransistor sealed in a single lightproof package.

Figure 5-12 shows how an opto-coupler can be used. When a voltage is applied across the + and – terminals, a current flows through the LED and it lights, illuminating the phototransistor and turning on the transistor. This brings the output down to close to 0V. When the LED is not powered, the transistor is off and R1 pulls the output up to 5V.

The important point is that the lefthand side of the circuit has no electrical connection to the righthand side. The link is purely optical.

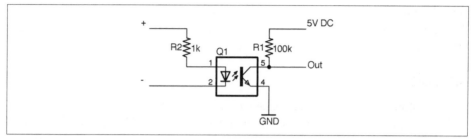

Figure 5-12. An Opto-Coupler

Discussion

When the sensing transistor is a TRIAC (see Recipe 5.6) the device is known as an opto-isolator rather than opto-coupler and the low-power TRIAC inside the opto-isolator can be used to switch AC using a more powerful TRIAC as shown in Figure 5-13. This circuit is often called an SSR (solid-state relay) as it performs much the same role as a relay would in AC switching, but without the need for any moving parts.

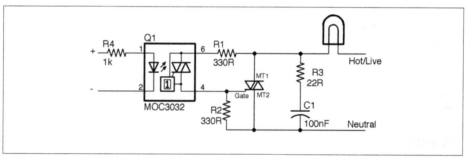

Figure 5-13. An SSR Design

Switching AC Is Dangerous

You should not work with high voltages unless you are completely sure that you can do so safely and understand the risks. See Recipe 21.12 for more information on working with high voltages.

R1 is used to limit the current flowing into the gate of the opto-isolator and R2 keeps the TRIAC MT1 turned off until the TRIAC in the opto-isolator is turned on by light falling on it because of the LED being energized. R3 and C1 form a filter to reduce the RF noise that would otherwise be generated by the switching. Note that R3 and C1 must both be rated to cope with the peak AC voltage (use 400V devices).

The MOC3032 package also contains a zero-crossing switch that stops the TRIAC from turning on until the AC crosses 0V. This greatly reduces the interference generated by the switching circuit.

See Also

For SSRs use opto-isolators, see Recipe 11.10.

For the MOS3032 datasheet, see *http://www.farnell.com/datasheets/2151740.pdf*.

For relays that also isolate the control side from the power side, see Recipe 6.4.

5.9 Discover Integrated Circuits

Problem

You want to know about ICs and how to use them.

Solution

ICs will be involved in almost every project you are likely to work on. Be it a microcontroller or special-purpose chip for motor control or a radio receiver or audio amplifier, there is probably a chip for it. There is little point in building a circuit out of transistors and other components if there is a single IC that will do the job. You will find many different sorts of IC scattered throughout this book.

ICs come in many different shapes and sizes. Figure 5-14 shows a selection of them.

Figure 5-14. A Selection of ICs

Discussion

ICs can have as few as three pins or hundreds of pins. You will often find that the same basic IC is available in different packages, both SMD and through-hole packages. This can be very useful if you plan to prototype your design using a solderless breadboard before manufacturing using SMD components.

Another way of using SMDs on a solderless breadboard is to use a "breakout board" that allows you to solder the SMD onto a board and then plug that board into the breadboard. There is an example of this shown in Figure 18-5.

ICs are identified by their part number, which is often difficult to read, so it's a good idea to label the bag the IC comes in.

ICs with more than three pins will have a little dot next to pin 1. If there is no dot, then there will be a notch that indicates the top of the IC. The pin numbering then moves down the IC to the bottom continuing with the bottom pin of the righthand side. For some example IC pinouts, see Appendix A.

See Also

You can find a list of the ICs used in this book in "Integrated Circuits" on page 414.

## CHAPTER 6

Switches and Relays

6.0 Introduction

Mechanical switches allow the flow of current to be turned on and off by flipping a toggle from one position to another or by pressing a button. They are so simple that they require little in the way of explanation other than to describe the various types of switches available and to highlight their limitations.

Relays long predate transistors as a means of using a small current to switch to a much bigger one. However, their flexibility and separation of the control side from the switching side mean that they are still in use today.

6.1 Switch Electricity Mechanically

Problem

You want to understand how a switch works.

Solution

Switches generally work by mechanically bringing together two metal contacts. Figure 6-1 shows how this might work.

When the push button is pressed, it presses the sprung metal contact at the top to the fixed contact in the base of the switch.

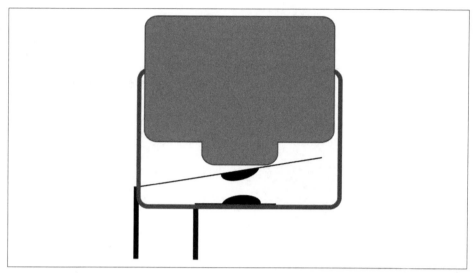

Figure 6-1. The Mechanical Construction of a Push Switch

Discussion

There are several things that can make switching less straightforward than you might expect:

1. When switching high voltages there may be arcing (sparking) just before the two contacts connect and even more so when they disconnect. This causes heat and can damage the contacts.
2. When switching high currents at any voltage, there is a danger of the contacts spot-welding themselves together as the contact is made, preventing the switch from turning off again.
3. A continuous high current through a switch will cause the contacts (which have a small resistance) to heat up, shortening the life of the switch.
4. The contacts often *bounce*, making and releasing contact a few times in very rapid succession before settling. This can cause a problem when used with software that toggles something on and off. In Recipe 12.1 you will see how a switch press can be *debounced*.

For these reasons switches have a maximum voltage and current, and there are often separate maximums for switching AC and DC.

See Also

For a discussion of the various types of switch, see Recipe 6.2.

6.2 Know Your Switches

Problem

You want to understand what types of switches are available and how to use them.

Solution

Figure 6-2 shows a selection of switches.

Figure 6-2. A Selection of Switches

The switches from left to right are:

- A tactile push switch. This is probably the most commonly used switch in commercial products. They are low-cost circuit board–mountable push switches.
- A panel mount push switch. These are good for creating one-off projects as they can just be mounted into a hole drilled into an enclosure.
- A slide switch. Moving the slider connects the center contact to either the left or right pins.
- Microswitch. The hinged lever and mounting holes make it easy to attach to a mechanical assembly. Because they have a hinged lever, they do not need much force to operate. These switches are durable and reliable and often built into microwave ovens to turn off the power if the door is opened.
- A toggle switch. These are the classic on/off switches.

Discussion

If you are designing a project, then you are probably most likely to be using a micro-controller of some sort, in which case a simple push switch is all that is required. Even if you have a situation where you might want to use a switch to select between two options, chances are you would use a pair of push switches and some feedback as to which is selected in a display.

Slide and toggle switches have mostly been relegated to the role of switching power on and off to the project, although you will occasionally find them on circuit boards as "selector" switches.

Toggles (and other switches) are available in many different switching configurations. Some of those described include DPDT, SPDT, SPST, and SPST. These letters stand for:

- D—Double
- S—Single
- P—Pole
- T—Throw

A DPDT switch is double pole, double throw. The word *pole* refers to the number of separate switch elements that are controlled from the one mechanical lever. So, a double-pole switch can switch two things on and off together. A single-throw switch can only open or close a contact (or two contacts if it is a double pole). However, a double-throw switch can make the common contact be connected to one of two other contacts. Figure 6-3 shows some of these configurations.

As seen in Figure 6-3 when drawing a schematic with a double-pole switch, it is common to draw the switch as two switches (S1a and S1b) and connect them with a dotted line to show that they are linked mechanically.

The matter is further complicated because you can have three poles or more on a switch, and double-throw switches are sometimes sprung, and do not stay in one or both of these positions. They may also have a center-off position where the common contact is not connected to any other contact.

Another kind of switch that you may encounter is the rotary switch. This has a knob that can be set to a number of positions. If you have a multimeter, then this is the kind of switch that you will use to select the range. Such switches are not often used. Rotary encoders coupled with a microcontroller (Recipe 12.2) are the more modern way of making a selection using a control knob that you turn.

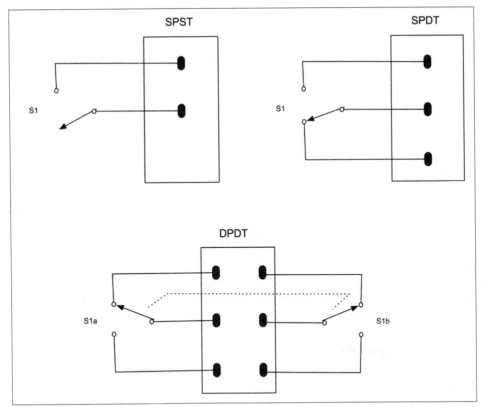

Figure 6-3. Switch Poles and Throws

See Also

To use switches with an Arduino or Raspberry Pi, see Recipe 12.1.

6.3 Switching Using Magnetism

Problem

You want to turn something on when a magnet is brought near.

Solution

A reed switch is made up of a pair of contacts running parallel to each other that are close but not touching. The contacts are enclosed in a glass capsule and when a magnet is brought close to the contacts they are drawn together, closing the switch.

Figure 4-6 shows a reed switch with and without a magnet next to it.

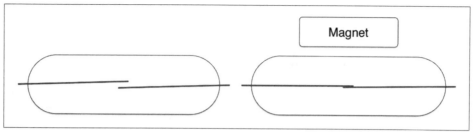

Figure 6-4. Switch Poles and Throws

Discussion

Reed switches are often found in intruder alarms on doors and windows. A fixed magnet is attached to the door and the reed switch to the door frame. When the door is opened, the magnet moves away from the reed switch, triggering the alarm.

See Also

Reed switches are also found in reed relays (*https://en.wikipedia.org/wiki/Reed_relay*) but are rarely used these days.

6.4 Rediscover Relays

Problem

You want to understand electromechanical relays and how to use them.

Solution

An electromechanical relay has two parts, a coil that acts as an electromagnet and switch contacts that close when the coil is energized. Figure 6-5 shows a relay, along with the pinout of the relay and a photo of an actual relay.

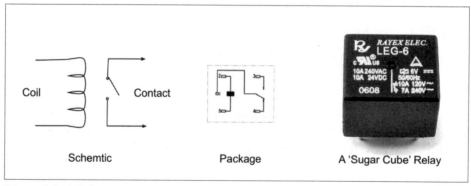

Figure 6-5. A Relay

Discussion

Although somewhat old-fashioned, relays are still in use today. They are equally happy switching AC or DC and are easily interfaced to a microcontroller.

If you take a relay apart, you will probably see something like Figure 6-6.

Figure 6-6. Inside a Relay

Here you can clearly see the coil that forms the electromagnet that when energized will move the contacts seen at the top.

The electromagnet of a relay is a big coil of wire and when released will produce a voltage spike that should be snubbed using a diode (see Recipe 11.9).

See Also

See Recipe 11.9 for information on using a relay with an Arduino or Raspberry Pi.

Today SSRs are often used in place of electromechanical relays (see Recipe 11.10).

## CHAPTER 7

Power Supplies

7.0 Introduction

Anything electronic needs a source of power. This may be as simple as a battery, but often will involve reducing high-voltage AC to the normal 1–12V DC that most electronics use.

Sometimes you need to generate higher voltages from a low-voltage battery. This may be stepping up the supply from a single 1.5V AA battery to a 6 or 9V, or it may be to generate much higher voltages for applications such as the 400V to 1.5kV supply needed by Geiger-Müller tubes.

The ultimate high-voltage power supply is a solid-state Tesla coil (see Recipe 7.15).

This is the first chapter where you will encounter a fair selection of ICs. When using an IC that you are not familiar with, your first port of call should be its datasheet. It will not only tell you how to avoid destroying it (by making sure you do not exceed its maximum ratings) but will also tell you how it behaves and if you are lucky will often include "reference designs" that are complete circuits that use the chip in a practical situation. These designs are developed by the chip manufacturer to show you how to use the IC, and many of the recipes in this chapter start from such reference designs.

If the datasheet for the IC does not include such useful designs, then the next thing to search for is "application notes" for the IC. This will often expand the rather terse and scientific-looking datasheet into practical circuits using the IC.

7.1 Convert AC to AC

Problem

You want to know how to convert AC at one voltage to another voltage.

Solution

Use a transformer (see Recipe 3.9).

When buying a transformer designed for 60Hz AC, you will often find the windings in an arrangement something like the one shown in Figure 7-1.

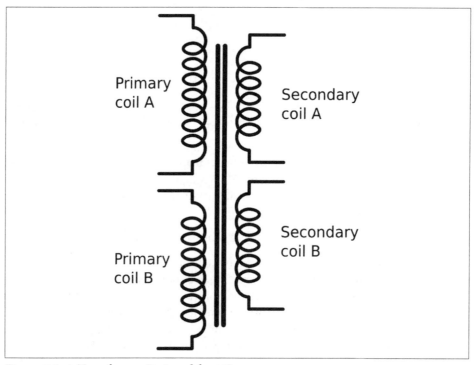

Figure 7-1. A Transformer Designed for AC

This has two identical primary coils and two identical secondary coils, all wound on the same former (laminated iron frame). This allows some flexibility in how the transformer is used. For example, US electricity is 110V whereas much of the world uses 220V AC. If you are designing a product and need the same low-voltage AC output whether the input is 100V or 220V, then two primary coils allow you to do this if you power the primaries in parallel for 110V and series for 220V.

The dual secondaries provide similar flexibility to wire in series for double the output voltage.

Discussion

Transformers are only rated to handle a certain amount of power. The resistance of the windings causes them to heat as current flows through them, and if things get too hot the insulation on the wire will break down and the transformer will fail.

A transformer will normally be rated as so many VA (volt amps). For most loads connected to a transformer 1VA is the same as 1Watt, but if large inductive loads like motors are being driven, then the current and voltage become out of phase and the apparent power will be lower than the VA rating.

See Also

For an introduction to transformers, see Recipe 3.9.

Reducing the AC voltage is often the first step in making an unregulated power supply, which is the topic of Recipe 7.2.

7.2 Convert AC to DC (Quick and Dirty)

Problem

You want to reduce AC to low-voltage DC, but it's OK if the DC voltage rises and falls a little depending on the load current.

Solution

Use an AC step-down transformer (see Recipe 7.1) and then rectify and smooth the output.

Figure 7-2 shows one schematic for the simplest method of achieving this.

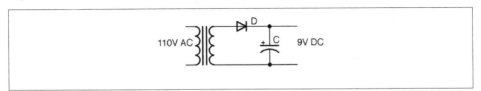

Figure 7-2. A Basic Unregulated AC-to-DC Power Supply

The output voltage in Figure 7-2 is shown as 9V. This will be 1.42 (square root of 2) of the stated output voltage of the transformer.

The diode D rectifies the low-voltage AC from the transformer (see Recipe 4.1). The capacitor C then smooths this out to a constant DC that theoretically will be the peak voltage from the cathode of the diode. The diode will prevent the capacitor from discharging back through the transformer and so once it is at the peak voltage it will stay there.

Discussion

The schematic of Figure 7-2 is missing any kind of load. The capacitor is being charged to the peak voltage, but nothing is using that charge. When you add a load to the output as shown in Figure 7-3 the capacitor will still be receiving its kicks of voltage from the diode and transformer, but now it will be discharging through the load at the same time.

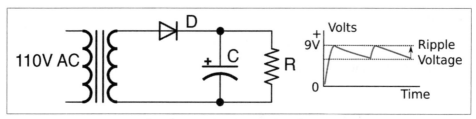

Figure 7-3. A Basic Unregulated AC-to-DC Power Supply

When using power supplies you will hear the use of the word "load," which is the thing that is being powered by the power supply. When thinking about how the power supply will operate, you can just think of this load as a resistance. In Figure 7-3 the load is represented by R.

The characteristic voltage drop during each cycle due to the load discharging C is called the ripple voltage.

The size of this ripple voltage can be calculated from the load current and also from the size of the capacitor according to this formula:

$$V_{ripple} = \frac{I}{2fC}$$

where I is the current in amps, f is the AC frequency (60Hz), and C is the capacitance in Farads.

As an example, if the load current is 100mA, a 1000μF capacitor would reduce the ripple voltage to:

$$V_{ripple} = \frac{I}{2fC} = \frac{0.1A}{2x60Hz \times 0.001F} = 0.833V$$

Notice that the ripple voltage is proportional to the load current, so in the preceding example, a current of 1A would result in a ripple voltage of 8.3V!

This state of affairs can be improved by using a full-wave rectifier as described in Recipe 7.3, which effectively doubles the frequency *f* in the preceding formula, halving the ripple voltage.

See Also

For full-wave rectification, see Recipe 7.3.

7.3 Convert AC to DC with Less Ripple

Problem

The half-wave rectification of Recipe 7.2 is not very efficient. You want to reduce the ripple voltage by using full-wave rectification.

Solution

There are two solutions to this:

- If you have a center-tapped or dual secondary transformer, you just need two diodes for full-wave rectification.
- If you only have a single secondary winding, you can use four diodes as a bridge rectifier.

Full-wave rectification with dual windings

Figure 7-4 shows how you can make use of both the positive and negative halves of the AC cycle.

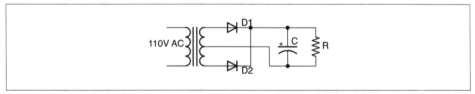

Figure 7-4. Full-wave Rectification using Dual Secondary Windings

The secondary windings are in serial, with the central connection between the windings becoming the ground of the DC output. When the end of the winding connected to D1 is at its maximum, the end of the winding connected to D2 will be at its minimum (greatest negative value). D1 and D2 will take it in turn to provide positive voltage to the capacitor and the whole AC cycle can contribute to charging the capacitor.

Figure 7-5 shows the full-wave rectified output before it is smoothed by the capacitor. It is as if the negative cycles are being reversed.

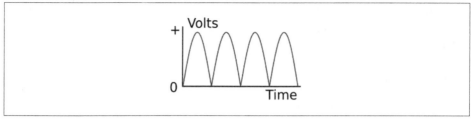

Figure 7-5. Full-wave Rectification

Using a Bridge Rectifier

If you only have a single secondary winding, then you can still achieve full-wave rectification by using an arrangement of four diodes called a bridge rectifier. Figure 7-6 shows an unregulated DC power supply using a bridge rectifier.

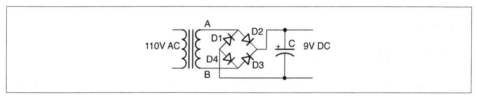

Figure 7-6. Using a bridge rectifier

To understand how this works, imagine that point A is positive and B is negative. This will allow current to flow from A through D2 to charge the capacitor. If there is a load on the power supply, then the current from the negative side of the DC output will be able to flow back through D4 to B.

When the AC cycle flips polarity and it is B that is positive and A negative, then the positive DC output is supplied through D3 and returns to A through D1.

Discussion

In Recipe 7.2 you saw how the ripple voltage halves if you use full-wave rectification rather than half-wave. The ripple voltage can be reduced by using very large capacitors, but in most circumstances it is better to use a regulated power supply as described in Recipe 7.4 or a switched mode power supply (SMPS), as discussed in Recipe 7.8.

You can buy a bridge rectifier as a single component that contains the four diodes wired up in a bridge configuration. This simplifies the wiring of your power supply as the bridge rectifier will have four terminals: two marked with a ~ to indicate AC input as well as + and − terminals for the rectified output.

See Also

To learn about half-wave rectification, see Recipe 4.1 and Recipe 7.2.

7.4 Convert AC to Regulated DC

Problem

You need a DC power supply that does not have any significant ripple voltage when under load.

Solution

Use a linear voltage regulator IC after your unregulated DC power supply.

Figure 7-7 shows such a circuit. As you can see, the first stage of the project is just an unregulated power supply.

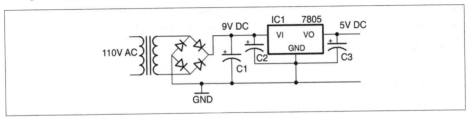

Figure 7-7. A Regulated DC Power Supply

A linear voltage regulator IC such as the very common 7805 regulator has three pins:

- GND
- Vin—unregulated DC input
- Vout—regulated DC output

In the case of the 7805, the output voltage is fixed at 5V, so as long as the input voltage is above about 7V the output will not have any significant ripple voltage (because the regulator chip needs about 2V more than its output voltage of 5V to successfully regulate the voltage).

At first sight, since C1 and C2 are in parallel, it is not immediately obvious why you would need both. However, the reason is that C1 will be a large electrolytic capacitor (470μF or more) and C2 will be a smaller MLC (typically 330nF) with a low ESR that should be positioned as close as possible to the voltage regulator. C2 along with C3, which is typically 100nF, are required by the voltage regulator IC to ensure that the regulation remains stable.

Discussion

In selecting values for C2 and C3, it is best to consult the datasheet for the voltage-regulator IC that you are using. This will normally specify values for the capacitors as well as tell you other important properties of the voltage regulator such as:

- Maximum current that can be drawn from the output (1A for a 7805)
- Maximum input voltage (35V for a 7805)
- Dropout voltage—the number of volts by which the input should exceed the output for normal operation (2V for a 7805)

Some linear voltage regulators are described as low-dropout (LDO). These have a much smaller dropout voltage than the 2V of the 7805. Some only need as little as 150mV more at the input than the output. If the input voltage of the regulator does fall below the dropout voltage, the output voltage will gradually decrease as the input voltage decreases. These regulators such as the LM2937, which has a dropout of 0.5V, are great if your input voltage is unavoidably close to the desired output. Perhaps you have a 6V battery and require a 5V output. Keeping the input voltage close to the output voltage also helps to reduce the heat generated by the regulator.

Voltage regulators are available in various packages from tiny surface-mount devices to larger packages such as the TO-220 package of the 7805, which are designed to be bolted to a heatsink for higher current use.

Most voltage regulators including the 7805 have protection circuitry that monitors the device's temperature and if it gets too hot, it drops the output voltage and therefore limits the output current so that the device is not destroyed by the heat. Voltage regulators also have a mounting hole designed to be attached to a heatsink. This is not needed for low currents, but as the current gets higher a heatsink (see Recipe 20.7) will become necessary.

The 05 part of the 7805 refers to the output voltage and, as you might expect, you can also find 7806, 7809, 7812, and other voltages up to 24V. There is also a parallel 79XX range of negative voltage regulators should you need to provide regulated positive and negative power supplies as you do for some analog applications.

The 78LXX range of regulators are a lower-power (and smaller package) version of the 78XX regulators.

See Also

You can find the datasheet for the 7805 here: *https://www.fairchildsemi.com/datasheets/LM/LM7805.pdf.*

These days, most regulated AC-to-DC power supplies use SMPS design (Recipe 7.8).

7.5 Converting AC to Variable DC

Problem

You need a regulated DC supply, but you want to be able to control the output voltage.

Solution

Use an adjustable voltage regulator like the LM317.

Figure 7-8 shows a typical schematic for a variable output voltage regulator using the LM317.

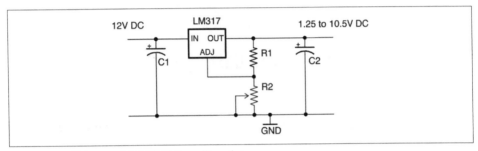

Figure 7-8. Using an LM317 Voltage Regulator

The output voltage of the LM317 varies according to the formula:

$$V_{out} = 1.25\left(1 + \frac{R_2}{R_1}\right)$$

This only holds true as long as R2 is relatively low resistance (less than 10kΩ).

So, if R1 is 270Ω and R2 is a 1kΩ pot, then at one end of the pot's travel, the output voltage will be:

$$V_{out} = 1.25\left(1 + \frac{R_2}{R_1}\right) = 1.25\left(1 + \frac{0}{270}\right) = 1.25V$$

When the pot's knob is rotated fully the other way, the output voltage will be:

$$V_{out} = 1.25\left(1 + \frac{R_2}{R_1}\right) = 1.25\left(1 + \frac{2000}{270}\right) = 10.5V$$

Discussion

As R2 increases, the output voltage also increases, but as with a fixed-output regulator, the regulator requires some spare voltage to do the regulating with. In the case of the LM317, this is about 1.5V less than the input voltage.

Adjustable voltage regulators like the LM317 are available in a number of different-sized packages and able to handle different amounts of current before their temperature-protection circuitry kicks in to prevent damage to the IC.

See Also

You can find the datasheet for the LM317 here: *http://www.ti.com/lit/ds/symlink/lm317.pdf*.

To use an LM317 as a current regulator, see Recipe 7.7.

7.6 Regulate Voltage from a Battery Source

Problem

You want a regulated fixed voltage (say 5V) from a battery-based power supply (say 9V).

Solution

Use a fixed linear voltage regulator as shown in Figure 7-9.

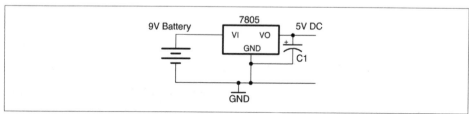

Figure 7-9. Using a Voltage Regulator with a Battery

Discussion

New batteries have a voltage that often starts a little higher than the rated voltage (perhaps 9.5V for a 9V battery). This voltage drops as the battery is used. A 9V battery will quickly fall to 8V and will probably only provide useful power until it falls to perhaps 7.5V as it's almost completely empty.

A voltage regulator can be used to provide constant output voltage during the lifetime of the battery and also to reduce voltage to a specific voltage needed for a microcontroller (usually 3.3V or 5V).

Unlike the transformer-based DC supply of Recipe 7.4 a battery does not suffer from ripple voltage, so the capacitor at the input to the voltage regulator can be omitted. However, the capacitor at the output is still needed in most applications as the load is likely to vary a lot and could cause instability in the regulator.

See Also

For more information on batteries, see Chapter 8.

7.7 Make a Constant-Current Power Supply

Problem

You need to provide constant current to a load, say, to a high-power LED.

Solution

Use an LM317 voltage regulator configured in constant-current mode as shown in Figure 7-10.

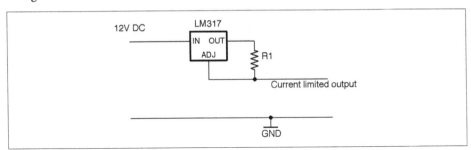

Figure 7-10. Using an LM317 as a Constant-Current Source

For the LM317 chip, the output current is set by the value of R1 using the formula:

$$I = \frac{1.2}{R_1}$$

So, to limit the current to a maximum of 100mA the resistor value would be chosen using:

$$R_1 = \frac{1.2}{I} = \frac{1.2}{0.1} = 12\Omega$$

Discussion

This circuit will automatically raise and lower the output voltage to keep the current at the desired level.

See Also

To see this circuit limit the power to a high-power LED, see Recipe 14.2.

7.8 Regulate DC Voltage Efficiently

Problem

You want to regulate your DC power supply in an energy-efficient manner that does not generate too much heat.

Solution

Use a switching voltage-regulator IC like the one in the schematic shown in Figure 7-11.

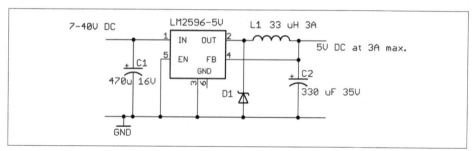

Figure 7-11. A Switching Voltage Regulator Using the LM2596

The LM2596 can provide a regulated power supply at 3A without needing a heatsink.

The FB (feedback) pin allows the regulator to monitor the output voltage and adjust the width of the pulses to keep the voltage constant. The EN (enable) pin can be used to turn the IC on and off.

Discussion

Linear regulators as described in Recipe 7.4, Recipe 7.5, and Recipe 7.6 suffer from the disadvantage that they simply burn off the excess voltage as heat, which means they get hot and also waste energy.

Switching regulator designs such as the buck regulator shown in Figure 7-11 operate at 85% efficiency almost independently of the input voltage. By comparison, a linear regulator with a high input voltage and a low output voltage may only have an efficiency of perhaps 20 to 60%.

In a similar manner to PWM (see Recipe 10.8) a buck converter uses a transistor to switch power to an inductor that stores the energy of each pulse at a high frequency (150kHz for the LM2596). The longer the pulse the higher the output voltage. A feedback mechanism changes the pulse length in response to the output voltage, ensuring a constant output.

It makes little sense to design a switching power supply from discrete components when there are ICs like the LM2596 that will do the whole thing so well.

There are many switching voltage-regulator ICs available and generally their datasheet will include a well-designed application circuit and sometimes even a circuit board layout.

If you are making a one-off project, then it is worth considering buying a ready-made switching-regulator module, either from eBay or suppliers like Adadruit and Sparkfun that specialize in modules.

See Also

You can find the datasheet for the LM2596 here: *http://bit.ly/2lOLtHc*.

7.9 Convert a Lower DC Voltage to a Higher DC Voltage

Problem

You have a low-voltage supply (say a single 1.5V cell) and want to boost the voltage to a higher voltage (say 5V).

Solution

Use a boost-converter IC as the basis for a design like the one shown in Figure 7-12.

This design will produce an output voltage of 5V from an input voltage of between 0.9V and 5V at 90% efficiency.

The SW (switch) output of the IC is used to send pulses of current through the inductor L1 generating high-voltage spikes that charge C1. The output VOUT is monitored through the FB (feedback) pin via the voltage divider formed by R1 and R2 that sets the output voltage.

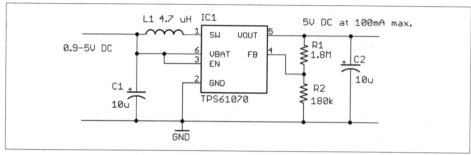

Figure 7-12. Using a TPS61070 Boost Converter

Discussion

The potential divider formed by R1 and R2 sets the output voltage. The value of R2 should be fixed at 180kΩ and the value of R1 calculated using the formula:

$$R_1 = 180k\Omega \times \left(\frac{V_{out}}{0.5V} - 1\right) = 180k\Omega \times \left(\frac{5V}{0.5V} - 1\right) = 1.62M\Omega \approx 1.8M\Omega$$

There are many similar boost-converter ICs available. In all cases, read the datasheet carefully as it will include a reference design and guides for finding the correct values for external components.

See Also

The datasheet for the TPS61070 can be found here: *http://www.ti.com/lit/ds/symlink/tps61070.pdf.*

7.10 Convert DC to AC

Problem

You want to convert low-voltage DC into high-voltage AC.

Solution

Use an inverter. These devices use an oscillator to drive a transformer and generate high-voltage AC from low-voltage DC.

High-Voltage AC

An inverter generates high-voltage AC, high enough voltage and current to stop your heart. So, do not attempt to make an inverter unless you are absolutely certain you have the knowledge and skills to do so safely.

Generating 110V from a low-voltage source (say a 12V car battery) is in theory very simple, but unless you really want to make such a thing yourself, it is generally cheaper and safer to buy a ready-made inverter product.

For more information on working with high voltages, see Recipe 21.12.

Discussion

If you really do wish to make yourself an inverter, Figure 7-13 shows a typical schematic.

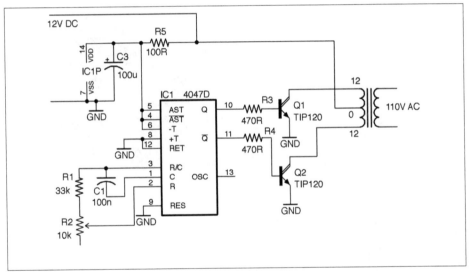

Figure 7-13. An Inverter Schematic

This design uses the CD4047 timer IC as an oscillator. This IC provides a similar function to the much more famous 555 timer (see Recipe 7.13, Recipe 7.14, and several recipes in Chapter 16) but has the advantage that it has both normal and inverted outputs. These two ouputs each drive an NPN Darlington power transistor to energize one half of the primary coil in turn.

The transformer used here is the kind used in recipes such as Recipe 7.1 to step down an AC voltage, but in this case, it will be used to step the voltage up. The windings that would normally be used as the secondary side of the transformer will be driven by the transistors and the output from the transformer will be taken from the coil that would normally have been connected to AC.

To drive the circuit from 12V DC at 60Hz to produce a 110V AC output you need a transformer with a 12-0-12V secondary and a 110V primary.

The power to the CD4047 is supplied through a 100Ω resistor (R5) and a 100μF decoupling capacitor (C3; see Recipe 15.1) across the IC's power supply. These reduce noise on the power supply to the CD4047.

Figure 7-14 shows the inverter built on a breadboard. This arrangement is fine as a first prototype, but in practice the design would be better built onto a circuit board and the power transistors attached to sizeable heatsinks (Recipe 20.7). It would also be wise to include a fuse in the 12V supply line to limit the current to a value that the transistors can handle, otherwise the transistors could overheat and die.

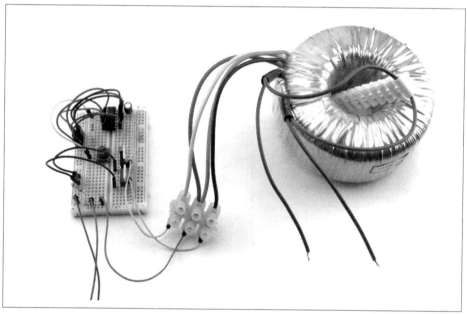

Figure 7-14. An Invertor Circuit Built on Breadboard

The transformer used is a toroidal type that I had available, but any type of transformer can be used.

See Also

For information on using a 4047 IC see its datasheet (*http://bit.ly/2mHK0Xk*).

To learn about fuses, see Recipe 7.16.

To build circuits using a solderless breadboard, see Recipe 20.1.

See Recipe 3.9 for more information on transformers.

7.11 Power a Project from 110 or 220V AC

Problem

You want to power your project from 110V or 220V AC supply efficiently, without using a large transformer like the one used in Recipe 7.2.

Solution

Designing SMPSs intended for high-voltage use is a specialized and potentially dangerous activity. Aside from the always present danger of electric shock, a poorly designed SMPS can easily overheat and become a fire hazard.

For these reasons, I strongly suggest that when you need to power a project from AC, use a ready-made commercial SMPS "wall wart" power adapter and fit a DC barrel jack socket to your project. These sealed adapters are readily available and made in such quantities that they will almost certainly be cheaper to buy than to make.

Discussion

It is interesting to see how these power supplies work. Figure 7-15 shows how such a power supply can achieve the same power output of a transformer-based design (Recipe 7.2) with perhaps $\frac{1}{10}$th of the weight.

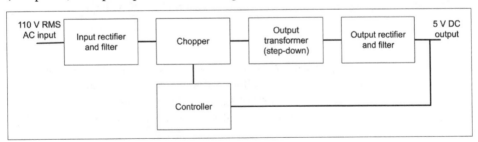

Figure 7-15. An SMPS

The high-voltage AC input is rectified and smoothed to produce high-voltage DC in the same way as Recipe 7.2 but without using a transformer first. This DC is then "chopped" or switched at high frequency (often around 60kHz) into a series of short pulses that then feed an output transformer that steps down the high-frequency AC. The resulting low-voltage AC is then rectified and filtered into low-voltage DC. Because the transformer operates at such a high frequency, it can be made much smaller and lighter than a 60Hz transformer.

Voltage regulation is achieved by feedback from the DC output to the controller that alters the width of the pulses being chopped that in turn alters the DC output voltage.

To keep the DC output isolated from the high-voltage AC, the feedback path to the controller uses an opto-isolator (Recipe 5.8).

This might seem like a lot of extra complexity, but the size, weight, and cost advantages of reducing the transformer size is enough to make SMPSs worthwhile.

Although there are ICs that do a lot of the work of an SMPS, they still require quite a few external components such as an opto-isolator and a high-frequency transformer.

Another advantage of SMPSs over transformer-based power supplies is that they can generally accept input voltages from 80–240V, and if the input voltage is higher, you just end up with shorter pulses to generate the same output voltage.

See Also

For the old-fashioned transformer-based approach to AC to low-voltage DC conversion, see Recipe 7.2.

For more information on transformers see Recipe 3.9 and for opto-isolators see Recipe 5.8.

7.12 Multiply Your Voltage

Problem

You have an AC voltage that you want to step up to a higher DC voltage.

Solution

This is a job for the very elegant voltage-multiplier circuit, which uses a ladder of diodes and capacitors to increase the voltage without the use of inductors. It can be built in multiple stages to increase the voltage by different multiples. The schematic of Figure 7-16 shows a four-stage voltage multiplier that will multiply the AC voltage by 4.

Discussion

To understand how this circuit works, consider what happens at the peak positive and negative value of the AC inputs. When the input is at its negative maximum, C1 will be charged through D1. On the next positive maximum, C1 will still be charged to the peak, but now, the input will be effectively added to C1. If we stopped there this would be a voltage doubler, but now C2 will charge from the next negative peak and so on until the output is at four times the peak input voltage.

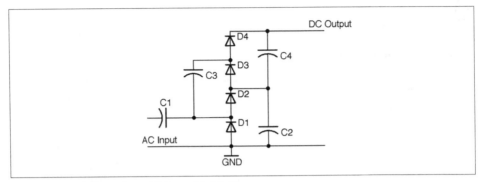

Figure 7-16. A Four-Stage Voltage Multiplier

The diodes should all be specified with a maximum voltage greater than the RMS AC input voltage times 1.4 (peak voltage). When selecting the capacitors (which should all be of the same value), you have a similar problem to that of selecting a smoothing capacitor in Recipe 7.2. If there is no load on the output then you can use very low value capacitors, but as soon as you add a load, they will discharge through it creating a ripple. Given that you will want the capacitors to be nonpolarized (no electrolytics here) and high-voltage, then you are likely to use something like 10nF capacitors rated for the same maximum voltage as the diodes for a low-current load such as a Geiger–Müller tube.

This kind of circuit is often used in high-voltage applications to multiply an initially high voltage up to even higher voltages.

See Also

For an example of using this kind of voltage multiplier in a power supply for a Geiger–Müller tube, see Recipe 7.14.

For background information on diodes and capacitors, see Recipe 4.1 and Recipe 3.1, respectively, and for background on AC see Recipe 1.7.

7.13 Supply High Voltage at 450V

Problem

You need a 450V DC low-current voltage source from a battery to supply power to a Geiger–Müller tube in a radiation meter.

Solution

Warning: High Voltage

This circuit can generate voltages approaching 1000V. Although this is at low current, it will still give you a nasty jolt and could have disastrous consequences if you have a pacemaker or heart problems, so please be careful if you decide to make this recipe.

This is especially true if you reuse the transformer from the flash unit of a disposable camera, as these units have a large capacitor charged to 400V or more that can really hurt. So discharge the capacitor first (Recipe 21.7).

Use a version of the inverter circuit of Recipe 7.10 to drive a small high-frequency transformer. Figure 7-17 shows the schematic for the power supply. Note that D1 and C4 should both be rated at 1000V.

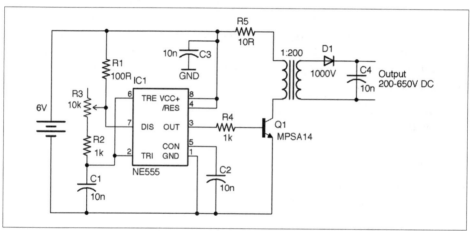

Figure 7-17. Schematic for High-Voltage DC Supply

The 555 timer is designed to provide a frequency of between 7kHz and 48kHz controlled using the variable resistor R3. The Darlington transistor provides pulsed current to the transformer, the output of which is rectified and smoothed.

Altering the frequency will alter the output voltage. With a high-voltage multimeter (1000V DC) adjust the variable resistor until the maximum voltage is found. For my transformer this took place at a frequency of 35kHz.

Discussion

Figure 7-18 shows the project built on two pieces of breadboard. The left breadboard has the 555 oscillator and the righthand board the transistor and high-voltage part of the design.

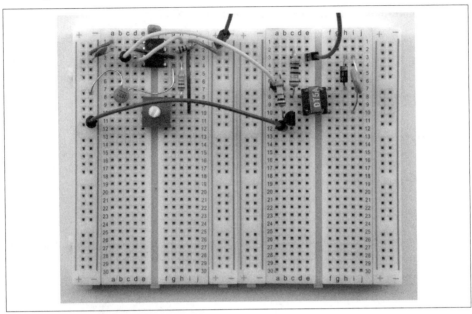

Figure 7-18. High-Voltage DC Supply on Breadboard

As you can see, the high-frequency transformer is tiny. The device I used is marked DT5A and is scavenged from the flash unit of an old disposable camera.

To check the output voltage, you will need a voltmeter with a 1000V DC range. Modern digital multimeters generally have an input impedance of around 10MΩ. At 1000V that means that a load current of 1kV / 10MΩ = 100nA. This does not sound like much, but for a low-current circuit like Figure 7-17, this loading may well reduce the voltage being measured. Increasing the value of C4, perhaps by putting a couple of capacitors of the same value in parallel with the original C4, should tell you if the multimeter is having an effect.

To measure high voltages reliably, you can use a specialist high-voltage meter with a very high input impedance, but these are expensive.

See Also

See Recipe 7.14 for a version of this design that adds a three-stage voltage multiplier to increase the output voltage.

For information on using the 555 timer, see Recipe 16.6.

For information on measuring high voltages, see Recipe 21.8.

To build circuits using a solderless breadboard, see Recipe 20.1.

7.14 Even Higher Voltage Supply (> 1kV)

Problem

The 450V output of Recipe 7.13 is not enough—you need 1.2–1.6kV for a Geiger–Müller tube that detects alpha radiation.

Solution

Add a three-stage voltage multiplier to the schematic of Recipe 7.13. The result of this is shown in Figure 7-19.

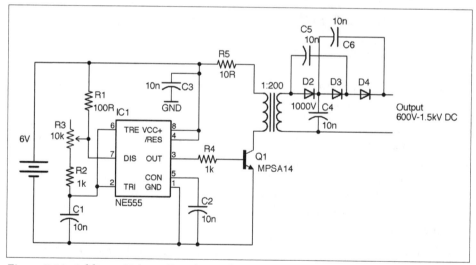

Figure 7-19. Adding a Voltage Trippler to the High-Voltage Supply

The diodes, C4, C5, and C6, can all still be rated at 1kV since the voltage across each should still be under 1kV.

Discussion

As with the design of Recipe 7.13 the output of this design is likely to contain a few volts of unwanted noise.

See Also

For a version of this design without the voltage multiplier, see Recipe 7.13.

7.15 Very Very High Voltage Supply (Solid-State Tesla Coil)

Problem

You want to make a Tesla coil.

Solution

One common and safe design for a Tesla coil uses a homemade transformer. Figure 7-20 shows the completed Tesla coil illuminating an LED from a distance of half an inch and Figure 7-21 shows the power supply built on the breadboard.

Figure 7-20. A Tesla Coil Lighting an LED

Figure 7-21. A Tesla Coil on Breadboard

Figure 7-22 shows the schematic for the project that uses a self-oscillating design based around a transformer and transistor.

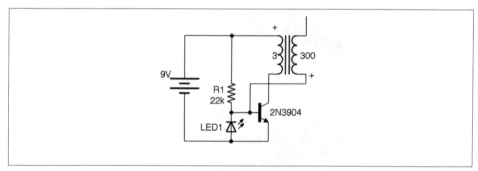

Figure 7-22. A Tesla Coil Schematic

The transformer for this design is made from a length of 2-inch plastic pipe onto which around 300 turns of 34 SWG (30 AWG) enameled copper wire are wound next to each other. The primary is made up of just three turns of plastic insulated multi-core wire.

This circuit is called a solid-state flyback converter (sometimes called a slayer exciter). The top end of the transformer secondary is not directly connected to anything, but with the top end of the coil connected to something with a large area (such as the copper ball I used) there is a small stray capacitance to ground. The top surface of the large area could be a metal plate or various other arrangements, as long as it's made from a large conducting area without too many sharp points to encourage discharge.

When power is first applied to the circuit, the transistor will turn on through R1, resulting in a large current flowing through the three turns of the transformer primary. This induces current to flow through the secondary, creating a potential difference due to the stray capacitance. Although small, this capacitance is sufficient to reduce the voltage on the bottom end of the secondary (and base of the transistor) to such an extent that the transistor turns off. The LED ensures that the base never falls more than 1.8V (the forward voltage of the LED) below ground. Because the transistor turns off, the magnetic field collapses and R1 reasserts its influence over the transistor's base to turn the transistor back on and so the cycle continues.

If you get the connections of the primary the wrong way around, or even just the geometry of the secondary wrong, there is a chance the transformer will turn on but oscillation fail to start, which will rapidly destroy the transistor. If all is well, LED1 will light. A lab power supply with current limiting is very useful in this situation (see Recipe 21.1).

Discussion

Holding a second LED close to the high-voltage end of the secondary will cause it to light. By holding one lead of the LED, you will be providing a weak path to ground.

A quick way to increase the power of the circuit is to place several transistors in parallel (all three pins). By using four 2N3904s I was able to generate enough voltage to cause the gas in a compact florescent lamp to ionize, lighting the lamp from a range of almost a foot.

See Also

For information on prototyping with a solderless breadboard, see Recipe 20.1.

You can see this Tesla coil in action at *https://youtu.be/-DEpQH7KMj4*.

For other types of high-voltage power supply, see Recipe 7.13 and Recipe 7.14. The "joule thief" circuit (Recipe 8.7) operates in a similar way to this design.

To see a miniature Tesla coil design in action using a ferrite core rather than an open tube, visit *https://www.youtube.com/watch?v=iMoDAspGPPc*. The design here uses the same schematic as in Figure 7-22.

If you search the internet for Tesla coil, you will find some incredible builds.

7.16 Blow a Fuse

Problem

You want to protect a circuit from too much current flowing, since too much current could destroy expensive components or cause a fire.

Solution

Use a fuse. Figure 7-23 shows a selection of fuses.

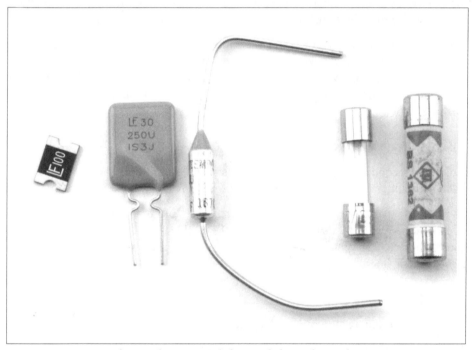

Figure 7-23. From Left to Right: SMD Polyfuse, Polyfuse, Thermal Fuse, 20mm Fuse, and 25mm Fuse

Fuses can be broadly classified as one-shot traditional fuses or polyfuses. As the name suggests, one-shot fuses can only be used once. When their current or temperature limits have been exceeded and they have blown, that's it, they must be discarded. For this reason they are generally attached to electronics using a fuse holder, so they can be replaced easily.

Polyfuses (a.k.a. resettable fuses) do not blow, but when the current through them increases above their limit value, their resistance increases, only resetting when the current returns to zero and they have cooled down. This means they can be permanently attached to a circuit. You will often find them used in places like USB ports, where you need to protect computer ports against overcurrent by misbehaving peripherals. Polyfuses use the same technology as PTC thermistors (see Recipe 2.9).

Discussion

In modern circuit design, and especially for low-voltage and relatively low-current use, polyfuses are the best approach. However, for AC applications and other high-current situations, where overcurrent will only happen if something is seriously wrong, then traditional single-use fuses are the safest option because they require manual intervention before the power can be reapplied to the downstream circuit.

If a single-use fuse blows, then work out why it blew before replacing it. The most likely cause of overcurrent is a short circuit, either from wires touching that should not be touching, or a component failing that causes some part of the circuit to draw more current than it should.

There are a number of different single-use fuses:

- Slow-blow—survives short periods of overcurrent. For example, for use with a motor that draws a lot of current when starting.
- Fast-blow—blows as soon as the current is exceeded.
- Thermal—designed to blow on overcurrent, but also if the fuse gets hot because of external influences such as fire.

See Also

To test a fuse, see Recipe 21.5.

For a circuit to produce a constant current, see Recipe 7.7.

To protect a circuit against accidental polarity reversal, see Recipe 7.17.

7.17 Protect from Polarity Errors

Problem

You are building a project that you want to be idiot-proof so that if an idiot puts the batteries in the wrong way, the project won't be harmed.

Solution

Many ICs and circuits designed with discrete transistors will be destroyed by overcurrent and hence heating if the power is applied with the wrong polarity.

If you don't care about a small voltage drop of 0.5V to 1V, then use a single diode from the positive battery terminal to the positive supply of your circuit (see Figure 7-24).

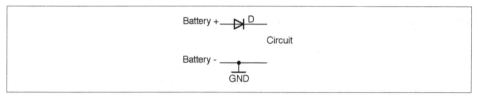

Figure 7-24. Protecting Against Reverse Polarity with a Diode

Remember to select a diode that can handle the current that you expect the circuit to draw.

If you can't afford to lose a volt through the diode but are losing 0.2–0.3V, then use a Schottky diode in place of the regular diode.

If even 0.3V is too much to lose, then use an P-channel MOSFET as shown in the circuit in Figure 7-25.

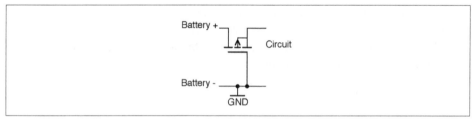

Figure 7-25. Reverse Polarity Protection Using a MOSFET

With the correct polarity, the P-channel MOSFET's gate-drain voltage will be high enough to keep the MOSFET on. Since MOSFETs with extremely low on-resistances are available, there will be very little voltage drop across the MOSFET to the load. The circuit only works for battery-supply voltages over the gate-threshold voltage, or the MOSFET will not turn on.

When the polarity is reversed, the MOSFET will be firmly off, preventing any significant current from flowing into the load circuit.

Remember that the link between the gate and other connections of the MOSFET is capacitative and so no current can leak through the gate to the negative terminal of the battery.

Although almost by tradition people tend to prefer to switch and use polarity protection on the positive supply, you can also do the same trick with an N-channel MOSFET on the negative battery supply as shown in Figure 7-26.

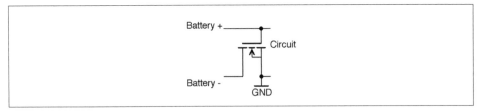

Figure 7-26. Negative Side Polarity Protection

Discussion

It's a good idea to include polarity protection in any battery-powered project, even with a 9V battery that has a clip that only goes one way around, because it is all too easy to touch the wrong contacts together.

Using a MOSFET is probably the cheapest way to protect a design, as it can cope with several amps and will generally be smaller and cheaper than a diode with the same properties, especially in surface-mount technology.

See Also

To use fuses to protect a circuit against overcurrent, see Recipe 7.16.

For background on diodes, see Chapter 4.

For more information on MOSFETs, see Recipe 5.3.

## CHAPTER 8

Batteries

8.0 Introduction

In Chapter 7 you learned about various ways of supplying low-voltage DC to your electronics designs from AC. In this chapter you will learn how to use various types of batteries and photovoltaic solar cells.

8.1 Estimating Battery Life

Problem

You want to know how long your battery is going to last.

Solution

The capacity of a battery is specified in Ah (ampere hours) or mAh (milliampere hours). To calculate how many hours your battery will last, you need to divide this number by the current consumption of your project in A or mA.

For example, a 9V PP3 rechargeable battery typically has a capacity of about 200mAh. If you were to connect an LED with a suitable series resistor that limited the current to 20mA, the battery should be good for about:

$$\frac{200mAh}{20mA} = 10A \ hours$$

Discussion

The estimate of time you get using the preceding calculation is very much an estimate and all sorts of factors such as the age of the battery, the temperature, and the current will affect the actual battery life.

If you are using a number of batteries in serial (e.g., in a battery holder that contains 4xAA batteries), then you do not get to multiply the battery life by 4, because the same current will be flowing out of each battery (Figure 8-1).

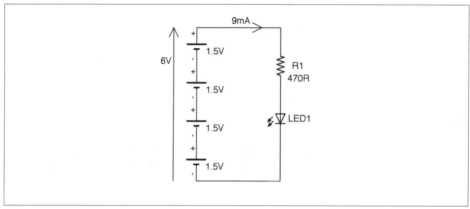

Figure 8-1. Batteries in Series

From Kirchhoff's Current Law (Recipe 1.4) we know the current flowing past any point in the circuit will be the same. Assuming an LED with a forward voltage of 1.8V the current will be around 9mA. Even though there is more than one battery cell, each cell will have 9mA flowing through it. A typical AA battery has a capacity of around 2000mAh and so we could expect this circuit to run for:

$$\frac{2000mAh}{9mA} = 222 hours$$

It is best to avoid putting batteries in parallel because none of the cells in the battery will be at exactly the same voltage and when you connect them together, the first thing that will happen is their voltages will equalize with the higher voltage batteries discharging into the lower voltage batteries. If the batteries are rechargeable, you may be okay if the voltage difference is very small, but for nonrechargeable batteries, this charging process can heat the batteries and be dangerous.

Batteries all have an internal resistance, very similar to the ESR of a capacitor (Recipe 3.2). This will cause heating as the battery discharges, which is why you cannot use a small 200mAh battery to supply 10A for 72 seconds. The internal resistance of the battery will both limit the current and cause the battery to heat up. Small batteries tend to have higher internal resistance than big batteries.

Most nonrechargeable batteries bought from a component supplier will specify the maximum continuous discharge current.

See Also

For information on choosing between different types of rechargeable batteries see Recipe 8.3 and for nonrechargeable batteries see Recipe 8.2.

To measure the current used by your project using a multimeter, see Recipe 21.4.

8.2 Selecting a Nonrechargeable Battery

Problem

You need a nonrechargeable battery for your project, but you are not sure what type to use.

Solution

Decide what your minimum battery life for the product should be and calculate the capacity of the battery you need in mAh, then use Table 8-1 to select a battery.

If you need more volts than a single battery cell provides, then place several of the cells in series in a battery holder.

Table 8-1. Nonrechargeable Battery Capacities and Voltages

Battery Type	Approx. Capacity (mAh)	Voltage (V)
Lithium Button Cell (e.g., CR2032)	200	3
Alkaline PP3 Battery	500	9
Lithium PP3	1,200	9
AAA Alkaline	800	1.5
AA Alkaline	2,000	1.5
C Alkaline	6,000	1.5
D Alkaline	15,000	1.5

Discussion

Higher capacity nonrechargeable batteries are expensive and so C and D cells are found less and less often in favor of rechargeable LiPo battery packs (see Recipe 8.3).

You should also be wary of using exotic or unusual batteries. It's generally a good idea to stick to universally available batteries like AA cells unless you need something smaller in which case AAA batteries will do.

Battery holders such as the 4 x AA battery holder shown in Figure 8-2 are readily available to hold 1, 2, 3, 4, 6, and 8 AA cells to make a battery pack capable of supplying 1.5, 3, 4.5, 6, 9, and 12V, respectively.

Figure 8-2. A 4 x AA Battery Pack

Perfect batteries would allow you to draw as large a current as you want. However, in reality batteries have a small resistance that will cause heating as the battery discharges. The higher the current the greater the heat produced. Smaller capacity batteries have a lower maximum discharge rate than larger batteries.

See Also

For information on rechargeable batteries, see Recipe 8.2.

8.3 Selecting a Rechargeable Battery

Problem

You want to select a rechargeable battery for your project, but are not sure what type to use.

Solution

Decide how long you would like your project to operate between charges and use Table 8-2 to choose a battery.

Table 8-2. Rechargeable Battery Capacities and Voltages

Battery Type	Approx. Capacity (mAh)	Voltage (V)
NiMh button cell pack	80	2.4-3.6
NiMh AAA cell	750	1.25
NiMh AA cell	2,000	1.25
LC18650 LiPo cell	2,000 to 8,000	3.7
LiPo Flat cell	50 to 8,000	3.7
Sealed lead acid (SLA)	600, 8,000	6 or 12

Discussion

LiPo (lithium polymer) and the closely related lithium ion cells are the lightest and also now as cheap per mAh as older technologies such as MiMh, but they require care when charging and discharging or they can catch fire (see Recipe 8.6).

You will still find NiMh batteries in products like electric toothbrushes. They are heavier than LiPos of the same capacity but are easy to charge (see Recipe 8.4).

SLA batteries (sealed lead acid) are even heavier than NiMh cells but cope well with trickle charging and last for years. However, their use is on the decline in favor of lighter-weight LiPo batteries.

Many rechargeable batteries will specify safe charge and discharge rates in units of C. This refers to a ration of the battery's nominal capacity. So, a 1Ah battery that says it can be discharged at 5C is capable of supplying 5A (5 x 1A). If the maximum charge current is specified as 2C the battery can be charged at 2A.

See Also

For a similar recipe but for nonrechargeable batteries, see Recipe 8.2.

8.4 Trickle Charging

Problem

You want to charge a NiMh or SLA battery while it's connected to your circuit and you don't need to charge it quickly.

 Don't Trickle Charge LiPo Batteries Like This

Note that LiPo batteries should not be trickle charged in this way. Instead, see Recipe 8.6 on charging LiPo batteries.

Solution

Trickle charge the battery from a power supply using a resistor to limit the current. If you don't know how the power supply works, then also include a diode in the circuit to prevent damage to the power supply when the power is turned off. Figure 8-3 shows the schematic for charging a 6V battery from a 12V supply.

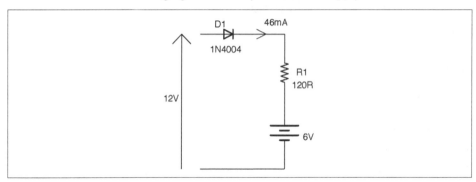

Figure 8-3. Trickle Charging

Discussion

A battery's datasheet will specify the charging current to use for fast charging and in this case slow trickle charging in terms of C, where in this case, C is not capacitance but rather the capacity of the battery in mAh. So the datasheet for an AA rechargeable battery might say that it should be trickle charged at C/10 mA, meaning that if the battery is a 2000mAh AA cell, it can be trickle charged at 2000/10 = 200mA indefinitely without damaging the battery. Charging below 200mA will still result in the battery being charged, and in situations like battery backup (Recipe 8.5) charging at C/50 (40mA) is probably a good compromise between charging speed and power consumption of the circuit.

To charge at 40mA we just need to set the value of R. The voltage across R will be 12V – 6V – 0.5V = 5.5V. We know that the current is 40mA, so we can calculate R using Ohm's Law as:

$$R = \frac{V}{I} = \frac{5.5V}{40mA} = 137.5\Omega \approx 120\Omega$$

Using the standard resistor value of 120Ω will result in a slightly higher current of:

$$I = \frac{V}{R} = \frac{5.5V}{120\Omega} \approx 46mA$$

We also need to make sure the resistor has a sufficient power rating, which we can calculate as:

$$P = IV = 46mA \times 5.5V = 253mW$$

So, a ¼W resistor is probably OK, but ½W would not get as hot.

An alternative to using a resistor to limit the current is to use Recipe 7.7 to provide a constant charging current. This design is more complex and expensive, but will cope better with variations in the input voltage and the battery's voltage changing as it charges.

See Also

Charging from a photovoltaic solar cell is very similar to trickle charging (see Recipe 9.1).

8.5 Automatic Battery Backup

Problem

You have a project that is powered through an AC power adapter but you want it to automatically change over to a battery if the AC power fails.

Solution

Use a battery of slightly lower voltage than the power adapter–supplied voltage and a pair of diodes in the arrangement shown in Figure 8-4.

The diodes ensure that the output is supplied from either the battery or the power supply, whichever has the higher voltage. Since you want the power supply to be providing the voltage most of the time, arrange for it to be a volt or more higher than the battery voltage (remember the battery voltage will drop as it discharges). In Figure 8-4, when the power supply is producing 10V, D1 will be forward-biased (conducting) and D2 reverse-biased. This reverse-biasing of D2 prevents any accidental charging of the battery (which may not be rechargeable). If the power supply were to turn off, D2 becomes forward-biased, providing the output voltage and D1 would become reverse-biased preventing any possible flow of current into the power supply that could damage it, especially if it is an SMPS design (Recipe 7.8).

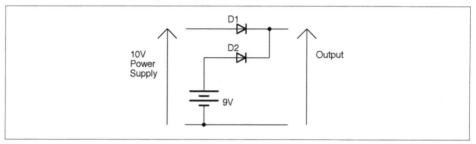

Figure 8-4. Automatic Battery Backup

Discussion

When selecting a diode type for D1 and D2, you need to make sure it can handle the current that the output needs to provide. Although in most cases regular diodes like the 1N4xxx series are just fine, they will result in a forward-voltage drop of at least 0.5V. You can reduce this voltage drop by using Schottky diodes (Recipe 4.2).

The circuit of Figure 8-4 will work just fine for rechargeable or nonrechargeable batteries, but it will not charge the batteries. If using rechargeable batteries, you can modify the circuit slightly so that the power supply trickle charges the battery and supplies the output. Figure 8-5 shows the schematic for this. See Recipe 8.4 for the calculation for the charging resistor.

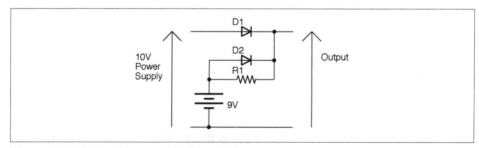

Figure 8-5. Trickle Charge and Battery Backup

Now when the power supply is in charge some of the current from D1 will flow through R1 and charge the battery (D2 is reverse-biased and can effectively be ignored). When the power supply is off (0V) D2 becomes forward-biased and supplies power to the output. At this time, R1 becomes irrelevant as D2 will be conducting.

See Also

See Recipe 4.2 for help selecting the right diode.

For the basics of what diodes do, see Recipe 4.1.

8.6 Charging LiPo Batteries

Problem

You want to charge a single LiPo cell.

Solution

Use a LiPo-charging IC designed specifically for the purpose of charging LiPo batteries. The schematic of Figure 8-6 is an example design using the MCP73831 IC.

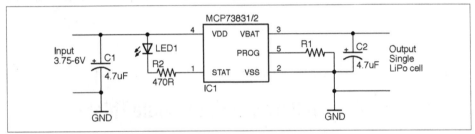

Figure 8-6. Charging a LiPo Battery

Many consumer electronics use LiPo batteries (often charged from a USB connection) with chips like the MCP73831 and many are low cost (less than $1) and include all sorts of advanced features that make sure the LiPo cell is charged safely.

The MCP73831 only needs two capacitors and a resistor to make a complete LiPo-charging circuit. The LED and R2 provide a status indicator that shows that the battery is charging and can be omitted if you don't need it.

The chip will automatically control the charge current, turning it to a safe trickle charge when the cell is fully charged. However, you can set the maximum charge current in amps (consult the datasheet for the LiPo cell that you plan to use) by means of R1, using the formula:

$$I_{max} = \frac{1,000}{R1}$$

So, to charge at 500mA (the maximum allowed by the chip anyway) set R1 to 2kΩ.

As an alternative to using a design of your own to charge a LiPo cell, you can also buy ready-made LiPo charge modules such as the Adafruit 1905 and Sparkfun PRT-10217 (both use the MCP73831).

Discussion

LiPo batteries have made the news in recent years because they will catch fire if abused. Using a charger chip like the MCP73831 in your circuits will help ensure your design is safe and reliable. For extra safety, you can add a thermal fuse (Recipe 7.16) to the supply to the charging circuit.

If you are looking for a lower-tech solution then consider using a NiMh or SLA battery instead.

Unlike NiMh or SLA batteries, LiPo batteries are not suited for charging in parallel. Each cell effectively has to be charged independently using special hardware. If your project cannot operate directly from the 3.7V of a LiPo cell and needs a bit more voltage, then it is generally easier to use a boost converter (Recipe 7.9) to generate the voltage you need.

See Also

You can find the datasheet for the MCP73831 here: *http://bit.ly/2n2XUPP*.

8.7 Get Every Drop of Power with the Joule Thief

Problem

You want to squeeze the last drop of energy from a battery.

Solution

The novelty circuit known as a joule thief allows you to power an LED from an alkaline AA cell even when its voltage drops to around 0.6V.

A joule thief is actually a small self-oscillating boost converter (see Recipe 7.9) that requires just a transistor, resistor, and homemade transformer to drive an LED from a single 1.5V battery, even when the battery voltage drops as low as 0.6V. The circuit is shown in Figure 8-7.

The circuit is very similar in operation to the self-oscillating boost converter used in Recipe 7.15 except that the secondary winding has the same number of turns as the primary and is connected back to the base via a resistor rather than using parasitic (from the winding) capacitance.

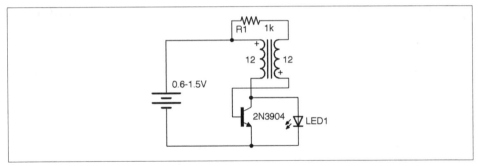

Figure 8-7. Joule Thief Schematic

Discussion

Figure 8-8 shows a joule thief. This circuit is interesting because it is fun to make one and see just how long an AA battery can power an LED.

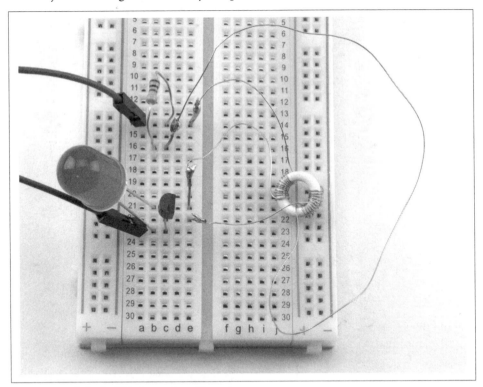

Figure 8-8. A Joule Thief on a Breadboard

I made my transformer by winding 12 turns of 30 AWG (34 SWG) enameled copper wire, for each winding, around a tiny toroidal ferrite core. In reality, more or less of any ferrite core and even plastic insulated wire will work just fine.

The supply generates pulsed DC at a frequency of about 50kHz (depending on the transformer).

See Also

For more practical boost converters, see Recipe 7.9.

Recipe 7.15 uses a very similar circuit to this recipe.

For information on wire gauges, see Recipe 2.10.

## CHAPTER 9

Solar Power

9.0 Introduction

This chapter is all about generating electricity from the sun using photovoltaic cells. The related topic of storing and using energy to power your projects or an Arduino or Raspberry Pi are also covered.

9.1 Power Your Projects with Solar

Problem

You want to understand how to power an electronics project using solar power, so that you are not dependent on power or the need to regularly replace batteries.

Solution

Use a photovoltaic solar panel (Figure 9-1) to generate electricity to trickle charge a battery that then supplies your project. The leftmost panel in Figure 9-1 is reclaimed from a low-power solar LED on a stick that cost about $1.50.

Figure 9-2 shows the schematic for solar charging a battery.

Solar panels are made up of a number of solar cells wired together in series to increase the output voltage. You will need a solar panel with an output voltage that is greater than the battery voltage.

The diode D1 is essential to prevent current from flowing back through the solar panel and damaging it when the solar panel is not illuminated enough for its output voltage to exceed the battery voltage.

R1 sets the charging current in just the same way as trickle charging (Recipe 8.4). As with trickle charging, you can also use a current regulator such as the one described in Recipe 7.7.

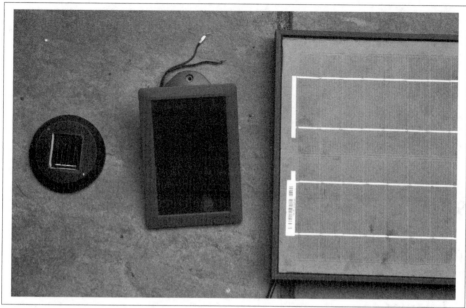

Figure 9-1. A Selection of Solar Panels—Left to Right Unknown Wattage, 1W, and 20W Panels

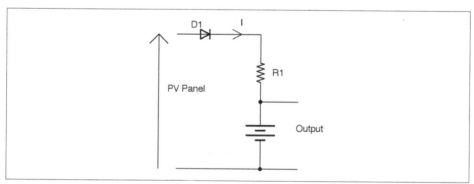

Figure 9-2. Solar Charging a Battery

Discussion

The performance of the solar panel varies enormously with the amount of sunlight falling on it. A solar panel will produce useful amounts of power when in direct sunlight, but in shadow or on a cloudy day it is unlikely to produce any more than 1/20 of its direct sunlight performance.

If you are thinking of using a solar panel to power a project indoors, unless you plan to keep the project on a sunny window sill or your project uses microamps, then forget it. The size of panel that you will need will make the project huge.

Two key parameters specified for a solar panel are its nominal output voltage and its power output. Both these values require some explanation.

The power output of the panel in mW for a small panel and W for a larger panel, as stated in the specification for the panel, is the power output under ideal conditions. That is, pointing directly at the sun without clouds, haziness, or any other impairments. In Table 9-1 you see the results of practical tests on the solar panels shown in Figure 9-1. These tests were carried out in the early afternoon at my home (53° latitude). The "no direct sunlight" readings were made with the panel pointing at clear sky but away from direct sunlight. For details on how to take such measurements yourself, see Recipe 9.3.

Table 9-1. Solar Panel Output Power

Nominal Power	Dimensions (inches)	Actual Output (direct sunlight)	Actual Output (no direct sunlight)	Guide Price (2016)
Unknown	1x1	40mW	1.4mW	$1
1W	6x4	210mW	8.4mW	$5
20W	22x12	6.9W	86mW	$40

Two things emerge from these measurements. First, the actual power output, even for the highest quality panel (the 20W panel), was actually almost ⅓ of its rated value. Secondly, when out of direct sunlight, the power output fell to almost 1/100 of the full-sunlight reading in the worst case. Now, while I'm sure the panels would do better at noon near the equator, that's not where most of us live.

Similarly, the output voltage at suboptimal lighting levels is also not the same as the specified output voltage. Table 9-2 shows the no-load output voltage for the same solar panels.

Table 9-2. Solar Panel Output Voltages

Nominal Output Voltage	Actual Output (direct sunlight)	Actual Output (no direct sunlight)
2V	2.2V	1.9V
6V	10.6V	9.7V
12V	21.3V	18V

The output voltage is higher than the nominal voltage so it can be used to charge a battery of nominal voltage.

See Also

To work out the power output of the solar panel that you need, see Recipe 9.3.

For the basics of trickle charging, see Recipe 8.4.

9.2 Choose a Solar Panel

Problem

You want to be able to decide on the output power of a solar panel to suit your project.

Solution

This solution will not give you an exact solar-panel power requirement and accompanying battery capacity, but it will at least give you a rough estimate to get you started with field testing.

Start by calculating the energy use of your project in Ws (joules) for a 24-hour period because a 24-hour day will include both the daylight and nighttime parts of the daily cycle. Calculating this current consumption is easy if the device is on all the time and the current consumption is constant. Let's say we have a garden temperature sensor that uses WiFi and consumes a constant 70mA from its 5V power supply. The power consumption is therefore 70mA x 5V = 350mW. The energy consumption is 350mW times the number of seconds in 24 hours. That is, 350mW x 24 x 60 x 60 = 30,240J.

This means that we need a solar panel that can during a 24-hour period provide 30,240J of energy. You can calculate the required solar-panel power using the formula:

$$P_{solar} = \frac{E}{H \times 24 \times 24 \times 60}$$

where E is the energy needed (30,240 in this example) and H is the average number of hours that the solar panel will be in direct sunlight. Let's assume that you live in the tropics, have more or less no seasonal variation in day length, and 10 hours of sunlight a day. In that case you can calculate the power of the panel you need as:

$$P_{solar} = \frac{E}{H \times 24 \times 24 \times 60} = \frac{30,240}{10 \times 60 \times 60} = 0.84W$$

So, in theory a 1W solar panel is fine for this particular situation.

If you think you might need the thermostat to keep working for three days without any solar charging (perhaps during a tropical storm) you will need a battery that can deliver 3 x 30,240 = 90,720J of energy.

The storage capacity in mAh (C) can be calculated using this formula:

$$C_{battery} = \frac{E}{V \times 60 \times 60} = \frac{90,720}{5 \times 3600} = 5Ah$$

Discussion

This equation has made a number of optimistic assumptions. First, the tropical day-light pattern is likely to be markedly different from a wet maritime climate at high altitude. In that case, you may only be able to assume a much lower average. For example, according to US Climate Data (*http://bit.ly/2mqgu7C*), December in Seattle only averages 62 hours of sunshine in the entire month (2 hours a day). This would require a solar panel of five times the power of the tropical equivalent.

The equation also assumes that any voltage conversion that is needed between the output of the solar panel down to the battery voltage is 100% efficient. If you are using a switching regulator like Recipe 7.8 then you might get 80%, but a linear regulator will not be as good, maybe only 50% or lower.

Also, the battery you are charging or the circuit charging it may limit the maximum power that you are using from the solar panel to prevent damage to the battery by charging it with too high a current. This all needs to be taken into account. Using a higher power solar panel than your battery can keep up with will have the advantage that under lower light conditions you will still get some charging.

All in all, using the worst-case sunshine hours scenario for your location, you should probably use double the calculated solar-panel output power requirement as your starting point, before ordering any solar panels to start experimenting with.

In addition to assumptions about the generation side of things, the load side is also often a lot more complicated than a constant load current. It may be that a device has a constant *standby* current most of the time, but then is turned on by a user or as a result of a timed operation and uses considerably more current for a while, before going back to standby mode. In addition to manually logging the current consumption and making estimates of the energy usage over 24 hours, you can also use a logging multimeter that will make and store periodic readings of current consumption. The resulting data can be downloaded into a spreadsheet for further analysis.

A logging multimeter can also be a great way of assessing the real performance of a solar panel (see Recipe 9.3).

If you are making a solar-powered project, then it is always worth working out how to keep the power consumption to a minimum. For an Arduino or other microcontroller project this may mean making the microcontroller go into a low-power sleep most of the time, just waking periodically to check if anything needs to happen.

Of course, some projects (maybe a solar-powered water pump) don't need a battery backup because they only need to operate when it's sunny.

See Also

For information on measuring the actual power output of a solar panel, see Recipe 9.3.

For examples of this approach for powering an Arduino see Recipe 9.4 and for a Raspberry Pi see Recipe 9.5.

9.3 Measure the Actual Output Power of a Solar Panel

Problem

Your solar panel has nominal output power, but you want to measure the actual output in your location.

Solution

Using a load resistor attached to the solar panel, measure the voltage across the resistor and from this calculate the power output. Figure 9-3 shows this arrangement.

The output power of the solar panel P is calculated as:

$$P = \frac{V^2}{R}$$

So, if you use a 100Ω load resistance and the voltage is 5V, the output power of the solar panel is:

$$P = \frac{V^2}{R} = \frac{25}{100} = 250mW$$

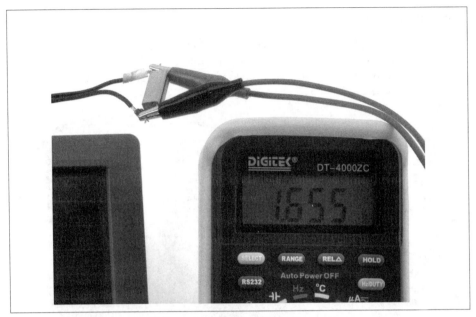

Figure 9-3. Measuring the Voltage Across a Load Resistor

The value of the load resistor needs to be low enough that the maximum output power of the solar panel is not exceeded, otherwise the solar panel will not generate as much as it could. In other words, if the solar panel is a 12V panel, you will probably find that the no-load output voltage is up to 18V in bright sunlight. So, you really want a load resistor value that gives an output of about 12V. You can calculate the maximum value of resistor to use using:

$$R_{load} = \frac{V_{nominal}^2}{P_{nominal}}$$

So, if your solar panel claims to be 20W at 12V, an ideal resistor value would be:

$$R_{load} = \frac{V_{nominal}^2}{P_{nominal}} = \frac{144}{20} = 7\Omega$$

Remember that the resistor will have to have a power rating sufficient to handle the full power output of the panel—in this case 20W.

Discussion

In addition to simply taking readings under different circumstances (full sunlight, shade, etc.) you can also use a logging multimeter such as the one shown in Figure 9-4 to record the voltage across the load resistor at regular intervals.

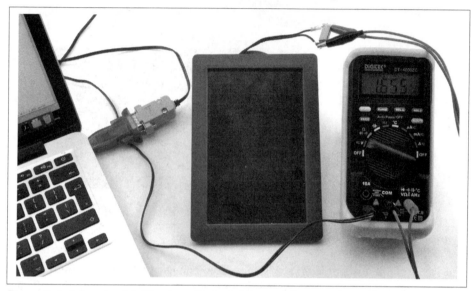

Figure 9-4. A Logging Multimeter with Serial Output Connected to a Computer

This way, you can create a picture of what your panel is likely to generate in a day.

See Also

In Recipe 9.2 this measurement technique was used to assess the performance of some of my solar panels.

For information on using a voltmeter, see Recipe 21.2.

9.4 Power an Arduino with Solar

Problem

You have an Arduino project that you want to power from a solar panel.

Solution

Use a 5V solar cellphone charger, like the one shown in Figure 9-5.

Figure 9-5. A Solar Charger/Battery

Discussion

Unless you are designing a product, using a ready-made charger circuit is by far the easiest way of solar powering your Arduino.

An Arduino Uno uses around 50mA, so a 4Ah charger/battery like the one shown in Figure 9-5 will power the Arduino for around:

$$\frac{4000}{50} = 80 hours$$

The Arduino Uno is not the most efficient of boards, when it comes to trying to minimize current consumption. It is more efficient to use an Arduino Pro Mini that has an external USB interface that you only need to connect when programming the Arduino. This can reduce your current consumption to 16mA. Using software to put the Arduino to sleep periodically can also greatly reduce current consumption.

See Also

For an introduction to Arduino, see Recipe 10.1.

To power your Raspberry Pi from solar, you will need a bigger solar panel than provided with ready-made phone boosters, as the Raspberry Pi is considerably more power hungry (see Recipe 9.5).

9.5 Power a Raspberry Pi with Solar

Problem

You want to power your Raspberry Pi from solar.

Solution

Use a 12V solar cell, charge controller, and SLA battery with a 12V to 5V power adapter.

This may sound excessive, but a Raspberry Pi with a WiFi adapter (WiFi is power hungry) can easily consume 600mA. If you add in a small 12V HDMI monitor, then you are likely to be drawing well over 1A.

Figure 9-6 shows a typical setup.

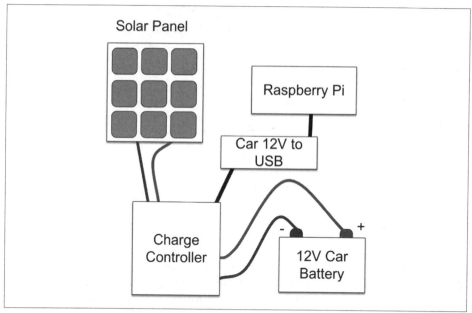

Figure 9-6. Solar Powering a Raspberry Pi

A ready-made charge controller generally has three pairs of screw terminals. One is attached to the solar panel, one to the battery, and one pair for the load, which in this case is a 12V to USB (5V) adapter commonly used as a phone car charger. Be sure to use one that can supply at least 1A.

Discussion

To run the Raspberry Pi continuously, you will need a 20W or greater solar panel. The choice of battery size depends on how long you want your Pi to keep running after it goes dark, or is dull for a few days. See Recipe 9.2 for information on how to make that choice.

If you are planning to use your Raspberry Pi as a controller, you may want to consider whether it would be better to use an Arduino, which uses only $\frac{1}{10}$ of the power, or if you need network connectivity, a Particle Photon or an ESP8266 module (see Recipe 10.6), both of which use less than 100mA.

See Also

To power an Arduino using solar energy, see Recipe 9.4.

## CHAPTER 10

Arduino and Raspberry Pi

10.0 Introduction

There is a good chance that any electronics project that you embark on will involve the use of a microcontroller (in the form of an Arduino) or a single-board computer (SBC) like the Raspberry Pi. Our gadgets are becoming more intelligent, requiring a little computer brain to control them; they are also becoming more connected, often needing an interface to the internet.

A typical Maker's electronics project these days will involve the use of a microcontroller or SBC with some electronics to switch things or sense value, or both. These extra electronics are attached to the microcontroller or SBC using its GPIO (general-purpose input/output) pins.

In this chapter the recipes mostly focus on the electronic side of interfacing with a microcontroller or SBC in general, but will also use the Arduino and Raspberry Pi as examples.

10.1 Explore Arduino

Problem

You want to understand just what an "Arduino" is and why it finds its way into so many electronics projects.

Solution

Figure 10-1 shows the most popular flavor of Arduino, the Arduino Uno R3.

Figure 10-1. An Arduino Uno

An Arduino is not a microcontroller but instead a microcontroller interface board. It has a microcontroller chip on the board, but it also has a whole load of other components that provide:

- Regulated power to the microcontroller
- A USB interface to program the Arduino from your computer
- A "power" LED
- A "user" LED connected to one of its pins that can be turned on and off programatically
- A 16MHz quartz crystal, necessary for the microcontroller's operation
- GPIO sockets for connecting external electronics

An Arduino does not do anything until it has been programmed. That is, you have to use your computer to write some instructions in the programming language C that can control and read the GPIO pins. Special software, called the Arduino IDE, is provided to both write the programs and then upload them onto the Arduino using a USB cable. Figure 10-2 shows the Arduino IDE with the "Blink" program loaded and ready to transfer to your Arduino. As its name suggests, the Blink program simply makes the Arduino's built-in LED blink on and off and is the starting point for most people's Arduino adventures.

What Programming Language Does Arduino Use?

Strictly speaking Arduinos are programmed using the C++ language, the "C" programming language with object-oriented extensions added to it. C is actually a subset of C++, which means that C programs that do not use any of the features of C++ will also work just fine. This means that most Arduino sketches (as Arduino programs are called) will look just like they were written in C rather than C++.

Because everyday writing of Arduino sketches does not require you to have any understanding of object-oriented software practices I prefer to say that Arduinos are programmed in C rather than C++, so as not to give the impression that you need to know anything about object-oriented programming to use an Arduino.

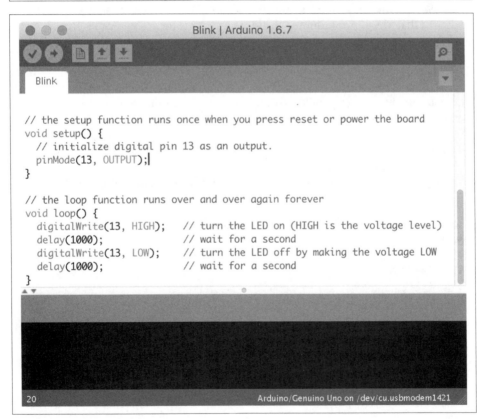

Figure 10-2. The Arduino IDE

You can download the Arduino IDE from *http://arduino.cc*.

In this book, it will be assumed that you have experiemented with Arduino and at least got as far as uploading a program onto it to make an LED blink. If you are new to programming or need a slower-paced introduction to Arduino, then see some of the book suggestions at the end of this recipe.

Discussion

In addition to the GPIO pins that you will learn more about in Recipe 10.7 the Arduino also has peripheral interfaces for I2C (Recipe 14.9 and Recipe 14.10) and SPI devices (Recipe 19.4).

The Arduino is an extremely useful and reliable component to use in your designs, even if it is just while you are prototyping the project and will later build your own design with a microcontroller.

In addition to the Arduino Uno shown in Figure 10-1, there are many other types of Arduinos that are all programmed the same way. This means you can select an Arduino based on its price, size, and number of GPIO connections.

See Also

There are many excellent books written about Arduino that will help you fill in the background. Two I particularly recommend are:

- If you are new to programming: *Programming Arduino: Getting Started with Sketches* by Simon Monk, TAB DIY, 2016.
- For an encyclopedic Arduino reference: *The Arduino Cookbook* by Michael Margolis, O'Reilly, 2011.

The following is a list of some of the Arduino-related recipes in this book, outside of this chapter:

- Recipe 11.6
- Most of Chapter 12
- Most of Chapter 13
- Most of Chapter 14
- Recipe 18.1
- Recipe 19.3
- Recipe 19.4

10.2 Downloading and Using the Book's Arduino Sketches

Problem

You don't want to have to type in all the example code in the book; you want to be able to download and use it.

Solution

All the Arduino sketches and Python programs for Raspberry Pi for this book are downloadable from GitHub (*https://github.com/simonmonk/electronics_cookbook*).

To use the Arduino sketches, download them from GitHub by cloning the directory if you are a git user, or use the Download Zip option behind the Clone or Download button on the GitHub page. You do not need to have a GitHub login to download the files.

Once extracted, the directory contains a directory called *arduino* with each of the Arduino sketches inside its own directory. Double-clicking on it will open the sketch in the Arduino IDE.

Installing the Arduino IDE

The Arduino IDE is available for Windows, OS X, and Linux. Before you can start programming your Arduino, you will need to download the IDE version for your computer's operating system following the instructions at the Arduino website (*https://www.arduino.cc/en/Main/Software*).

Discussion

An alternative way to access the sketches is to copy the contents of the *arduino* directory that you downloaded from GitHub into your Arduino IDE's sketches directory, which is found in your operating system's *Documents* or *My Documents* (Windows) folder in a directory called *Arduino*. Then you will be able to open any of the sketches from the File→Sketchbook menu of the Arduino IDE.

See Also

To download the Python files for Raspberry Pi, see Recipe 10.4.

If you are looking for a primer to teach you Arduino C, then my book *Programming Arduino: Getting Started with Sketches* (TAB DIY, 2016) should help you out.

10.3 Explore Raspberry Pi

Problem

You want to understand just what a "Raspberry Pi" is and why it finds its way into so many electronics projects.

Solution

A Raspberry Pi (Figure 10-3) is an SBC running a version of Debian Linux (called Raspbian) as an operating system. You can plug in a keyboard, mouse, and monitor and use it to browse the internet just like a "normal" computer.

Figure 10-3. From Left to Right: Raspberry Pi Zero Model A and Pi 2 Model B

The Raspberry Pi is available in a number of different sizes, from the very low-cost Pi Zero up to the Model 3, which includes built-in WiFi.

The reason the Raspberry Pi has become popular for electronics projects is that, like the Arduino, it also has GPIO pins that you can connect to external electronics. What's more, because the Raspberry Pi can be easily connected to the internet, you can also use it for all sorts of Internet of Things (IoT) projects.

When it comes to programming the Raspberry Pi, there are lots of options. In fact, all major programming languages are available to run on the Raspberry Pi, but the most popular language for electronics projects is probably Python, which is used in conjunction with the RPi.GPIO library.

While you need a computer to program an Arduino, to write programs for Raspberry Pi, you will use the Raspberry Pi itself.

Discussion

To use a Raspberry Pi unattended, you will want to be able to make your program automatically run when the Raspberry Pi Boots up (Recipe 10.3).

See Also

My book *The Raspberry Pi Cookbook* by O'Reilly (2016) is a cookbook like this, but dedicated entirely to the Raspberry Pi.

If you are new to programming and need a gentle introduction to Python programming on the Raspberry Pi, you might want to consider my book *Programming the Raspberry Pi: Getting Started with Python* (TAB DIY, 2015).

Here is a list of some of the Raspberry Pi-related recipes in this book outside of this chapter:

- Recipe 11.7
- Most of Chapter 12
- Most of Chapter 13
- Much of Chapter 14
- Recipe 18.2
- Recipe 19.2

10.4 Downloading and Running This Book's Python Programs

Problem

You want to get the Python programs for the Raspberry Pi in this book onto your Raspberry Pi so that you can run them.

Solution

All the Python programs for Raspberry Pi for this book can be found here: *https://github.com/simonmonk/electronics_cookbook*.

To use the programs, you can fetch them from GitHub directly onto your Raspberry Pi using the command:

```
$ git clone https://github.com/simonmonk/electronics_cookbook
```

This will also fetch all the Arduino sketches, but these can be ignored. You will find the Python programs in a directory called *pi*.

To run a program (e.g., *blink.py*) use the following command:

```
$ sudo python blink.py
```

Note that the sudo command is not required on the latest version of the Raspberry Pi's operating system (Raspbian) so you may be able to just type:

```
$ python blink.py
```

Discussion

Although I suggested ignoring the Arduino sketches earlier in this recipe, in reality you can download and install the Arduino IDE onto a Raspberry Pi and then use the Raspberry Pi to program the Arduino (see *http://spellfoundry.com/sleepy-pi/setting-arduino-ide-raspbian/*).

See Also

To access the Arduino code, see Recipe 10.2.

To set a Raspberry Pi Python program to automatically run on startup, see Recipe 10.5.

If you are looking for a primer to teach you Python with Raspberry Pi, my book *Programming Raspberry Pi: Getting Started with Python* (TAB DIY, 2015) should help you out.

10.5 Run a Program on Your Raspberry Pi on Startup

Problem

You want a program or script to start automatically as your Raspberry Pi boots.

Solution

Modify your *rc.local* file to run the program you want.

Edit the file */etc/rc.local* by using the command:

```
$ sudo nano /etc/rc.local
```

Add the following line after the first block of comment lines that begin with #:

```
/usr/bin/python /home/pi/my_program.py &
```

It is important to include the & on the end of the command line so that it runs in the background; otherwise, your Raspberry Pi will not boot.

Discussion

Be careful when you edit *rc.local*, or you may stop your Raspberry Pi from booting.

See Also

For general background on Raspberry Pi, see Recipe 10.3.

10.6 Explore Alternatives to Arduino and Raspberry Pi

Problem

Neither an Arduino nor Raspberry Pi are quite what you are looking for and you want to know what the alternatives are.

Solution

Table 10-1 lists some other popular boards (MC, microcontroller; SBC, single-board computer).

Table 10-1. Arduino and Raspberry Pi Alternatives

Board	Type	Notes	Arduino IDE Compatible	Website
Digispark	MC	A tiny Arduino compatible with just a few GPIO pins that plugs straight into your computer's USB port for programming.	Y	digistump.com
Adafruit Feather	MC	A small Arduino compatible with built-in LiPo battery charger and a range of wireless options.	Y	adafruit.com
NodeMCU	MC	A small, very low-cost board that can be easily converted to use the Arduino IDE. It has built-in WiFi.	Y	eBay
Particle Photon	MC	A small, low-cost Arduino compatible with built-in WiFi and IoT software built-in.	N	particle.io
Teensy3	MC	A small, low-cost Arduino compatible.	Y	pjrc.com
BeagleBone Black	SBC	A Raspberry Pi alternative with more GPIO pins and analog inputs.	N/A	beagleboard.org
ODROID-XU4	SBC	An 8-core 2GHZ South Korean monster SBC.	N/A	hardkernel.com

Some of these boards are shown in Figure 10-4.

Figure 10-4. From Left to Right: Digispark, Photon, NodeMCU, and BeagleBone Black

Discussion

The Arduino IDE is a very flexible piece of software that can be easily extended to operate with unofficial Arduino-type boards. This means that you can stick to the Arduino IDE and the Arduino C programming language to program Arduino-like boards that may use a different processor or include useful addons such as WiFi, Bluetooth, or battery-charging hardware.

The Particle Photon deserves special mention, since although it uses a language that is based on Arduino C, you do not program it with the Arduino IDE, but rather with a web-based IDE. Deployment is then over the internet, allowing remote updating of applications. It also includes a very easy-to-use software framework for building IoT projects.

Other good boards to use where WiFi is required are any of the boards based on the ESP8266 modules, such as the NodeMCU and smaller modules like the ESP01. These can be programmed from the Arduino IDE. The ESP8266 boards are not as simple to use as the Photon but are extremely cheap.

See Also

For information on the Arduino see Recipe 10.1 and for the Raspberry Pi, see Recipe 10.3.

10.7 Switch Things On and Off

Problem

You want to use the Arduino, Raspberry Pi, and other microcontrollers and SBCs to control external electronic components.

Solution

Microcontrollers and SBCs have GPIO pins that allow you to connect external electronics to them. Figure 10-5 shows the schematic for a typical GPIO pin within a microcontroller chip or SoC of a Raspberry Pi. The pin can function as either a digital input or digital output, under software control.

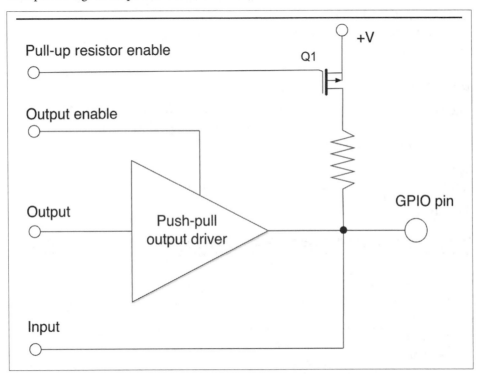

Figure 10-5. Schematic for a GPIO Pin

Referring to Figure 10-5, if the output enable control of the GPIO is set using software to be enabled, the pin functions as a digital output and the push-pull driver (Recipe 11.8) allows the GPIO pin to source or sink a few tens of milliamps.

If the push-pull driver is disabled, then the pin can function as a digital input. If the pull-up resistor enable control is set, then the transistor Q1 is enabled to allow the resistor to pull up the input to a default high. This is commonly used when connecting a switch to a digital input to prevent a floating input from oscillating between high and low (see Recipe 10.7).

Discussion

The GPIO schematic of Figure 10-5 is typical of most Arduino pins. Some of the Arduino pins (marked A0 to A5) can be used as analog inputs and for those pins, the GPIO is also connected to ADC hardware inside the microcontroller chip.

Some GPIO pins have pull-down resistors as well as pull-up resistors, which can be turned on and off in the same way as pull-down resistors.

Table 10-1 compares the GPIO features of an Arduino and Raspberry Pi 3.

Feature	Arduino Uno R3	Raspberry Pi 3
Operating voltage	5V	3.3V
Maximum individual output current	40mA	18mA
Maximum total current for all pins used as outputs	400mA	Not specified
Internal pull-up resistors	Y	Y
Internal pull-down resistors	N	Y
Number of GPIOs	18	26
Analog inputs	6	None

Figures 10-6 and 10-7 describe each of the accessible GPIO and power pins available on an Arduino and Raspberry Pi 3, respectively.

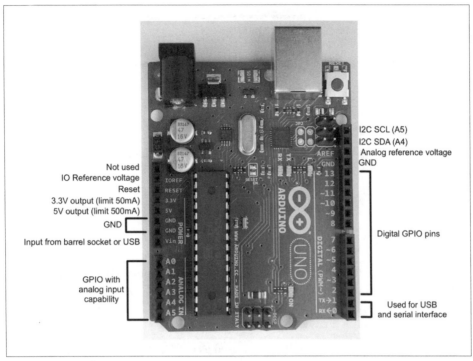

Figure 10-6. Arduino Uno R3 Pinout

Some of the pins require a little more explanation:

- IO reference voltage is the output voltage of the Arduino (5V for an Uno) but some other types of Arduinos operate at 3.3V. Intended for use by plug-in shields but rarely used.
- Vin is the supply voltage, which might be 9V if an external power supply is connected to the barrel jack, or 5V if the board is USB powered.
- The two I2Cs (IC to IC buses) are used for connecting to I2C devices (see Recipe 14.9), but are also connected to A4 and A5 of an Arduino Uno. On some other Arduino boards such as the Leonardo, these are separate.
- Analog reference voltage can be connected to a reference voltage below 5V to narrow the analog input range for greater precision at low voltages. If no connection is made, the analog inputs will be relative to 5V on an Arduino Uno.
- Pins 0 and 1 can in a pinch be used as extra GPIO pins, but if so you will need to disconnect anything attached to them to allow the USB connection to work. So generally it's best not to use them.

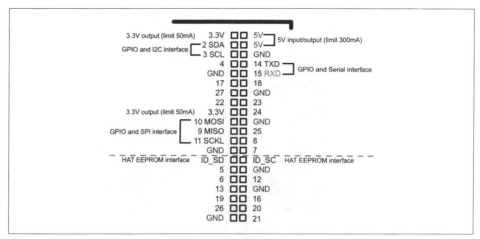

Figure 10-7. Raspberry Pi GPIO Connector Pinout

Unlike an Arduino, the GPIO pins on a Raspberry Pi are not labeled. If you are going to be connecting things to the GPIO pins, it's a good idea to get a hold of a GPIO template, such as the Raspberry Leaf (Adafruit 2196), that fits over the GPIO pins allowing you to identify them easily.

Most of the pins on a Raspberry Pi can be used as GPIO, but some have second functions:

- Pins 2 and 3 can also be used to connect I2C devices.
- GPIO pins 9 to 11 can also be used as an SPI for devices that support that connection type.
- ID_SD and ID_SC are dedicated to an interface for any HAT (hardware attached to top) that fits over the GPIO connector and allows software to identify the HAT.
- Pins 14 and 15 can be used to provide TTL serial interfaces for devices such as GPS modules that often use that interface standard.

If you have an older Raspberry Pi (before the model B+) it will only have 26 pins on the GPIO connector. These are the same as the top 26 pins (above the dashed line) of the 40-pin layout of newer Raspberry Pis shown in Figure 10-7.

See Also

For an introduction to Arduino see Recipe 10.1 and for Raspberry Pi see Recipe 10.3.

Many of the recipes in Chapters 11, 12, 13, and 14 use GPIO pins.

10.8 Control Digital Outputs with Arduino

Problem

You want to configure an Arduino GPIO pin to be an output and then turn the output on and off using software.

Solution

Use the pinMode function to set the pin to be an output and then use digitalWrite to turn the pin on and off. The following example program will make digital pin 13 (attached to the Arduino's built-in LED) blink on and off:

```
const int ledPin = 13;

void setup()
{
  pinMode(ledPin, OUTPUT);
}

void loop() {
  digitalWrite(ledPin, HIGH);  // turn the LED on
  delay(1000);                 // wait for a second
  digitalWrite(ledPin, LOW);   // turn the LED off
  delay(1000);                 // wait for a second
}
```

The code for this sketch is available as part of the book downloads (Recipe 10.2). The sketch is called *blink*.

Discussion

The program (or "sketch" as they are known in the Arduino world) starts by defining a constant for the pin connected to the LED called ledPin, which is given the value 13. If you change your mind and decide to make another pin blink, you only need to change 13 to that pin number in one place in the code.

The setup function is run just once after the Arduino is reset. The pinMode function then specifies that ledPin is to be an OUTPUT. It is perfectly possible to change a pin's mode while the sketch is running. You can find a good example of this with Charlie-plexing (see Recipe 14.6).

The loop function will be run repeatedly over and over again, and each time it is run, it will first set ledPin (pin 13) high, wait a second (1000 milliseconds) then set it low, then wait another second, and so on.

See Also

For digital outputs on a Raspberry Pi, see Recipe 10.9 and for digital inputs on an Arduino see Recipe 10.10.

For the current-handling capabilities of an Arduino digital output, see Recipe 10.7.

10.9 Control Digital Outputs from Raspberry Pi

Problem

You want to configure a Raspberry Pi GPIO pin to be an output and then turn the output on and off using software.

Solution

Use the RPi.GPIO library (included in Raspbian) with Python. The following example program will make GPIO pin 18 turn on and off once a second:

```
import RPi.GPIO as GPIO
import time

GPIO.setmode(GPIO.BCM)

led_pin = 18

GPIO.setup(led_pin, GPIO.OUT)


try:
    while True:
        GPIO.output(led_pin, True)   # LED on
        time.sleep(1)                # delay 1 second
        GPIO.output(led_pin, False)  # LED off
        time.sleep(1)                # delay 1 second
finally:
    print("Cleaning up")
    GPIO.cleanup()
```

The code for this program is available as part of the book downloads (see Recipe 10.4). The file is called *blink.py*.

Unlike the Arduino, the Raspberry Pi does not have any user-controllable LEDs built in so to see this program in action see Recipe 14.1.

Discussion

The code starts by importing the GPIO and time libraries. It then sets the GPIO pin identification mode to be BCM (Broadcom). This is a hangover from the early days of Raspberry Pi when two ways of identifying the pins were almost equally used. However, for historical reasons, you still need to include this line at the top of all your Python programs. Some internet and book resources still exist that use the pin positions on the connector rather than the pins' name. If you find such a resource, it will tell you to use a pin mode of BOARD instead of BCM.

The variable led_pin refers to the GPIO pin to be blinked and the pin set to be an output.

The main program loop is contained inside a try/finally block. This is not strictly speaking essential and the program will run just fine without it, but by calling GPIO.cleanup() whenever the program exits, all the GPIO pins are put back into a safe input state so that accidental shorts of the pins will not cause any damage.

Inside the while loop, the pin is first turned on, then there is a delay of a second, then it is turned off, etc. The sleep function takes a time in seconds as a parameter that can also be less than a second using a decimal notation. For example, to delay for half a second you would use time.sleep(0.5).

See Also

The Arduino equivalent to this program can be found in Recipe 10.8.

For digital inputs on a Raspberry Pi, see Recipe 10.11.

10.10 Connect Arduino to Digital Inputs Like Switches

Problem

You want to read an Arduino digital input in an Arduino sketch.

Solution

Use the Arduino C digitalRead function. To see the result of the read, use the Arduino Serial Monitor. The following sketch illustrates this:

```
const int inputPin = 7;

void setup()
{
  pinMode(inputPin, INPUT);
  Serial.begin(9600);
}
```

```
void loop()
{
  int reading = digitalRead(inputPin);
  Serial.println(reading);
  delay(500);
}
```

You can find this sketch in the downloads for the book (see Recipe 10.2). It is called *ch_10_digital_input*.

The constant inputPin is defined as pin 7 and initialized to be an INPUT in the setup function.

The loop function first assigns the value resulting from carrying out a digitalRead on inputPin to the variable reading and then sends this value over the Arduino's USB interface to the Serial Monitor. Finally, a delay of 500 milliseconds is added to slow things down to a manageable rate.

To open the Serial Monitor (Figure 10-8), click the rightmost icon on the Arduino IDE's toolbar. This looks like a magnifying glass.

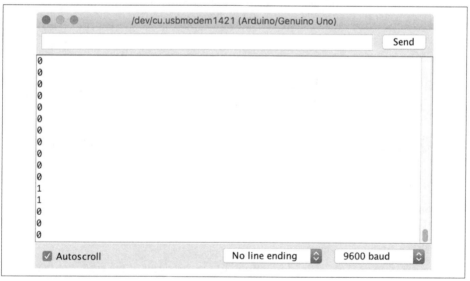

Figure 10-8. The Arduino Serial Monitor

You should see a steady stream of numbers appear as shown in Figure 10-8. These should be predominantly 0s, but there may be occasional 1s. Try attaching a male-to-male jumper wire or short length of solid core wire to pin 7 of the Arduino (Figure 10-9). You should find that you get a selection of 1s and 0s. This is because you are acting as an antenna and the digital input is picking up electromagnetic noise

---

(most likely "hum"). The digital input is "floating" and since it has a very high input impedance, it is very sensitive to noise.

Figure 10-9. A Floating Digital Input

Next, try attaching the floating end of the lead to one of the GND connections on the Arduino as shown in Figure 10-10. You should see that the output in the Serial Monitor is all 0s.

Figure 10-10. Connecting a Digital Input to GND

Finally, move the end of the wire that goes to GND to 5V and the Serial Monitor output should now be all 1s.

Discussion

The Arduino Serial Monitor is one of the few ways to see what is going on inside your Arduino and is often used to see what's going wrong in a sketch.

See Also

For digital inputs on a Raspberry Pi, see Recipe 10.11.

10.11 Connect Raspberry Pi to Digital Inputs Like Switches

Problem

You want to be able to read a GPIO pin as a digital input from your Python program.

Solution

Use the RPi.GPIO library. The following example program reads GPIO23 every half second and prints out the result:

```
import RPi.GPIO as GPIO
import time

GPIO.setmode(GPIO.BCM)

input_pin = 23

GPIO.setup(input_pin, GPIO.IN)

try:
    while True:
        reading = GPIO.input(input_pin)
        print(reading)
        time.sleep(0.5)
finally:
    print("Cleaning up")
    GPIO.cleanup()
```

The code for this program is available as part of the book downloads (see Recipe 10.4). The file is called *ch_10_digital_input.py*.

Discussion

When you run the preceding program, you will see mostly 0s in the terminal:

```
$ sudo python ch_10_digital_input.py
0
0
0
0
0
0
0
```

With both this program and Recipe 10.10 you can connect a switch to the digital input (Recipe 12.1) or use a female-to-female jumper wire to connect GPIO23 to either GND or 3.3V.

Do *Not* Connect to 5V

The maximum voltage for a GPIO pin on the Raspberry Pi is 3.3V. So on no account should you connect GPIO23 or any other GPIO pin to one of the 5V pins on the GPIO connector. Doing so is very likely to damage your Raspberry Pi.

See Also

For digital inputs on the Arduino, see Recipe 10.10.

10.12 Read Analog Inputs on Arduino

Problem

You want to read the voltage at a GPIO pin from your Arduino sketch.

Solution

Use the Arduino `analogRead` function and one of the Arduino pins A0 to A5 that can be used as analog inputs. The following example sketch writes the analog reading as a voltage to the Serial Monitor every half a second:

```
void setup()
{
  pinMode(inputPin, INPUT);
  Serial.begin(9600);
}

void loop()
```

```
{
  int reading = analogRead(inputPin);
  float volts = reading / 204.6;
  Serial.println(volts);
  delay(500);
}
```

You can find this sketch, which is called *ch_10_analog_input*, with the downloads (Recipe 10.2).

The sketch defines the inputPin as A0. When referring to one of the A0 to A5 pins on an Arduino, you have to use the letter and the number (A and 0), whereas for the other pins you just use the number.

The analogRead function returns a number between 0 and 1023, where 0 means 0V and 1023 means 5V. This number is converted to a voltage by dividing it by 204.6 (1023 / 5). Note that the analog range can be changed by connecting the Arduino AREF (analog reference) pin to a different voltage.

When you open the Serial Monitor you will see a series of values displayed. As with the digital inputs of Recipe 10.10 the analog input is floating and so the numbers (voltage on A0) will probably fluctuate something like this:

```
2.42
2.36
2.27
2.13
1.99
1.86
1.74
1.62
1.40
0.70
```

Discussion

Try the experiments in Recipe 10.11 and put a wire into A0. Touching the end of the wire with your fingers so that you act as a radio antenna should make the analog readings even more wild. Connecting A0 to the 5V connector of the Arduino should display a voltage reading of 5.00 on the Serial Monitor and GND will give you 0.00; connecting A0 to the 3.3V socket on the Arduino should give you a reading of around 3.30.

If the maximum voltage you want to measure is below 5V, then you can use the Arduino AREF pin to keep the 0–1023 range of readings but for a smaller voltage range and thus achieve higher precision.

For example, if you tie the AREF pin to the 3.3V pin on the Arduino you will get a full range of readings over the 3.3V range. The voltage at AREF needs to be stable and regulated or your readings will be inaccurate.

Why 0 to 1023?

The number 1023 may seem like a strange choice especially since it is so close to 1000, but it is used because it is 1 less than 2 to the power of 10 (2 multiplied by itself 10 times) and the analog value is digitized to 10 binary digits (bits).

See Also

The Raspberry Pi does not have analog inputs, but you can use an ADC IC (Recipe 12.4).

10.13 Generate Analog Output on Arduino

Problem

You want your Arduino to control the output power from a GPIO pin, say to control the brightness of an LED or the speed of a motor.

Solution

Use the `analogWrite` function of the Arduino on one of the PWM-capable pins.

The following example sketch will set the brightness of an LED wired to pin 11. The brightness is set by sending a number between 0 and 255 from the Arduino Serial Monitor to the Arduino board. For the curious, 255 is 1 less than 2 to the power of 8.

```
const int outputPin = 11;

void setup()
{
  pinMode(outputPin, OUTPUT);
  Serial.begin(9600);
  Serial.println("Enter brightness 0 to 255");
}

void loop()
{
  if (Serial.available())
  {
    int brightness = Serial.parseInt();
    if (brightness >= 0 && brightness <= 255)
    {
```

```
      analogWrite(outputPin, brightness);
      Serial.println("Changed.");
    }
    else
    {
      Serial.println("0 to 255");
    }
  }
}
```

You can find this sketch, which is called *ch_10_analog_output*, with the downloads for the book (Recipe 10.2).

In addition to illustrating PWM output from a GPIO pin, this example sketch also illustrates how you can send data from your computer to the Arduino via the Serial Monitor.

The setup function sets outputPin to be an OUTPUT and then starts serial communication. Finally, the setup function sends a message to the Serial Monitor instructing you how to use the sketch by sending a number between 0 and 255.

Inside the loop function, the call to Serial.available() tests to see if anything has been sent from the Serial Monitor, and if it has, it is converted into an int and assigned to the variable brightness. The analogWrite command is then used to set the output of outputPin.

To see this sketch in operation, you will first need to follow Recipe 14.1 and attach an LED to pin 11.

Open the Arduino IDE's Serial Monitor (see Recipe 10.10) and then try some different values between 0 and 255 to see how the brightness changes (Figure 10-11).

Not Working?

If, when you type a value into the Serial Monitor and click Send, the LED changes brightness momentarily, but then turns off, the Line ending drop-down list at the bottom of the Serial Monitor (see Figure 10-11) is probably not set to "No line ending."

What is happening is that the number is being sent and works, but then the line ending arrives as another message and is interpreted as being 0, turning the LED off again.

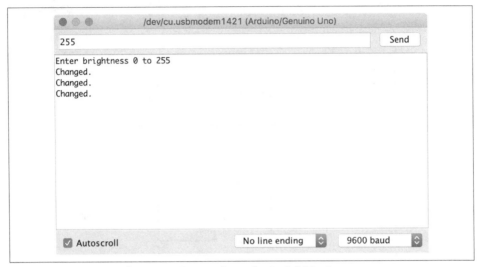

Figure 10-11. Setting the PWM Output from the Serial Monitor

Discussion

Not all the pins on an Arduino can be used for PWM like this. In fact, only the pins on the Arduino Uno marker with a ~ can. On an Arduino Uno, that's pins 3, 5, 6, 9, 10, and 11. Other Arduino models have different sets of pins that are PWM-capable, but you will need to refer to the Arduino documentation (*http://arduino.cc*) to know which ones.

Pulse-Width Modulation

In varying the brightness of an LED, it is tempting to think of PWM analog outputs as being truly analog and varying the output voltage. But this is not actually the case.

Figure 10-12 shows what is really going on.

The output pin is still acting digitally. That is, it's either logical high (5V on Arduino, 3.3V on Raspberry Pi) or it's low (0V). The duration of the pulses that emanate from the pin controls the brightness of an LED attached to a pin. Short pulses make the brightness low, and long pulses make it brighter on average and appear brighter to the observer.

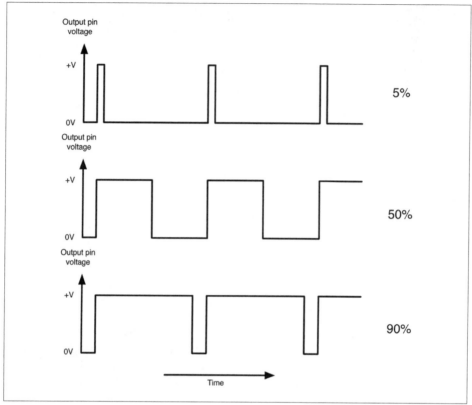

Figure 10-12. Pulse-Width Modulation

See Also

For the Raspberry Pi version of this recipe, see Recipe 10.14.

10.14 Generate Analog Output on Raspberry Pi

Problem

You want your Raspberry Pi to control the output power from a GPIO pin, say to control the brightness of an LED or the speed of a motor.

Solution

Use the PWM feature of the RPi.GPIO library to control the output power of a GPIO pin. The following Python example program illustrates this:

```
import RPi.GPIO as GPIO
```

```
led_pin = 18
GPIO.setmode(GPIO.BCM)
GPIO.setup(led_pin, GPIO.OUT)

pwm_led = GPIO.PWM(led_pin, 500)
pwm_led.start(100)

try:
  while True:
    duty_s = raw_input("Enter Brightness (0 to 100):")
    duty = int(duty_s)
    pwm_led.ChangeDutyCycle(duty)

finally:
  print("Cleaning up")
  GPIO.cleanup()
```

You can find this program in the downloads for the book (see Recipe 10.4). If you are using Python 3 rather than Python 2, change the command raw_input to just input.

To see this program in action, you will need to attach an LED to GPIO pin 18 (see Recipe 14.1).

The RPi.GPIO library is a little more complex to use than its Arduino counterpart. After defining the pin as an output, you then have to create a PWM channel using the line:

```
pwm_led = GPIO.PWM(led_pin, 500)
```

The number 500 is the frequency of the PWM pulses in Hz. This PWM channel is then started using the following line:

```
pwm_led.start(100)
```

Here, the value of 100 is the initial duty cycle (percentage of time the pin is high) of the PWM signal (in this case, 100% of the time).

The rest of the script interacts with the user requesting a value of the duty cycle between 0 and 100. Run the program and try different versions of brightness as shown here:

```
$ sudo python led_brightness.py
Enter Brightness (0 to 100):0
Enter Brightness (0 to 100):20
Enter Brightness (0 to 100):10
Enter Brightness (0 to 100):5
Enter Brightness (0 to 100):1
Enter Brightness (0 to 100):90
```

When you want to exit the program, press CTRL-C.

Discussion

The Raspberry Pi does not use a real-time operating system. That is, at any one time, there are many differnt processes running. So if you are trying to generate pulses of an exact length such as with PWM, you will find that you get a certain amount of jitter in the LED brightness as the pulse generation is interrupted.

See Also

For the Arduino counterpart to this recipe, see Recipe 10.13.

10.15 Connect Raspberry Pi to I2C Devices

Problem

You want to enable the I2C bus on your Raspberry Pi to connect I2C peripherals to it such as the displays used in Recipe 14.9 and Recipe 14.10.

Solution

In the latest versions of Raspbian, enabling I2C (and for that matter SPI—Recipe 10.16) is simply a matter of using the Raspberry Pi configuration tool that you will find on the main menu under Preferences (Figure 10-13). Just check the box for I2C and click OK. You will be prompted to restart.

On older versions of Raspbian, the `raspi-config` tool does the same job.

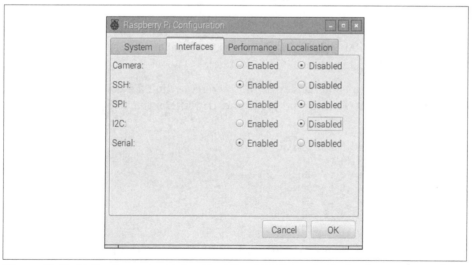

Figure 10-13. Pi Configuration Before Enabling I2C Using the Pi Configuration Tool

Start `raspi-config` using the following command:

```
$ sudo raspi-config
```

Then select Advanced from the menu and scroll down to I2C (Figure 10-14).

When you are asked "Would you like the ARM I2C interface to be enabled?" say "Yes." You will also be asked if you want the I2C module to load at startup, to which you should also say "Yes."

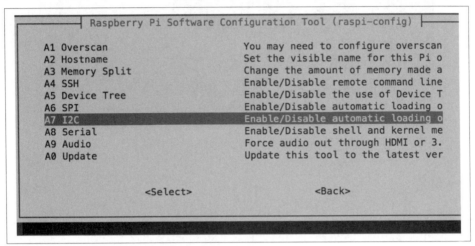

```
┤ Raspberry Pi Software Configuration Tool (raspi-config) ├

   A1 Overscan              You may need to configure overscan
   A2 Hostname              Set the visible name for this Pi o
   A3 Memory Split          Change the amount of memory made a
   A4 SSH                   Enable/Disable remote command line
   A5 Device Tree           Enable/Disable the use of Device T
   A6 SPI                   Enable/Disable automatic loading o
   A7 I2C                   Enable/Disable automatic loading o
   A8 Serial                Enable/Disable shell and kernel me
   A9 Audio                 Force audio out through HDMI or 3.
   A0 Update                Update this tool to the latest ver


             <Select>                      <Back>
```

Figure 10-14. Enabling I2C Using raspi-config

Discussion

I2C is a commonly used standard for connecting devices together. It uses two data pins SDA (data) and SCL (clock) to transfer data bidirectionally between devices. Usually one of the devices on the bus is a microcontroller or in this case, the SoC of the Raspberry Pi. Multiple devices can be connected to the same I2C bus pins, so, for example, you could have both a display and a sensor connected to the same two bus pins, each device having its own unique address.

To use I2C devices from Python, install the Python I2C library by using the commands:

```
$ sudo apt-get update
$ sudo apt-get install python-smbus
```

You will then need to reboot the Raspberry Pi for the changes to take effect.

When using I2C hardware the i2c-tools software can be a great help in debugging and making sure devices are properly connected to the Raspberry Pi. This can be installed using the command:

```
$ sudo apt-get install i2c-tools
```

When you have a device attached to the I2C bus, running the i2cdetect utility will tell you if it's connected and what I2C address it's using, as shown in Figure 10-15.

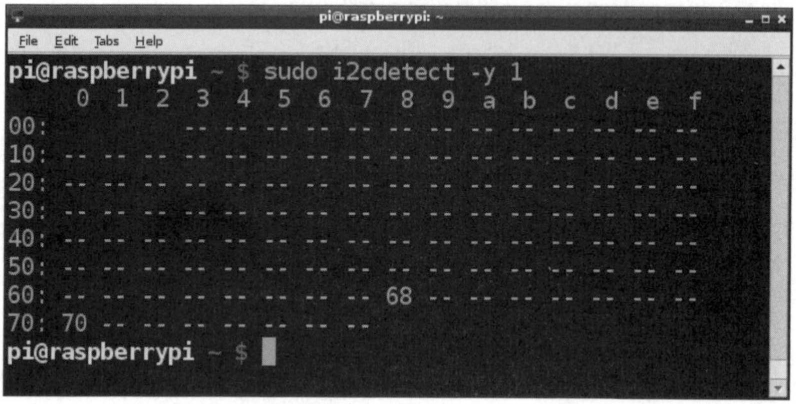

Figure 10-15. Using i2cdetect

See Also

To set up the Raspberry Pi's SPI, see Recipe 10.16.

For recipes that connect I2C peripherals to a Raspberry Pi, see Recipe 14.9, Recipe 14.10, and Recipe 19.3.

10.16 Connect Raspberry Pi to SPI Devices

Problem

You want to enable the SPI bus on your Raspberry Pi to connect peripherals to it.

Solution

By default, Raspbian is not configured for the Raspberry Pi's SPI. To enable it use the Raspberry Pi configuration tool found in the main menu under Preferences (as in Recipe 10.15), or on older versions of Raspbian, use `raspi-config` using the command:

```
$ sudo raspi-config
```

Then select Advanced, followed by SPI, and then "Yes" before rebooting your Raspberry Pi. After the reboot, SPI will be available.

Discussion

The SPI allows serial transfer of data between the Raspberry Pi and peripheral devices, such as ADCs and port expander chips, among other devices. It is similar in concept to I2C but uses four pins instead of two. As with I2C SPI data is synchronized with a clock signal (SCLK) but separate lines are used for each direction of communication: MOSI (master out slave in) and MISO (master in slave out) and a separate "Enable" pin is needed from the master for each of the devices connected to the bus. It is an older and less elegant standard than I2C but still widely used.

You may come across examples of interfacing with SPI that use an approach called bit banging, where the RPi.GPIO library is used to interface with the four GPIO pins used by the SPI.

See Also

Recipe 12.4 uses an SPI ADC converter chip.

The SPI is used in Recipe 12.4 and Recipe 19.4.

10.17 Level Conversion

Problem

You want to connect a 5V device to a Raspberry Pi or one of the 3.3V Arduino models.

Solution

When you are connecting a 3.3V-level output to a 5V input, with very few exceptions (see discussion), no level conversion is needed—you can connect the two directly.

However, if you want to connect a 5V output to a 3.3V input, then that's a different matter. Direct connection will wall damage the 3.3V device. If the 3.3V device is specified as having 5V-tolerant inputs, then you can just connect the two together directly. Note that the inputs of a Raspberry Pi are *not* 5V tolerant, so you should connect them together using a voltage divider as shown in Figure 10-16.

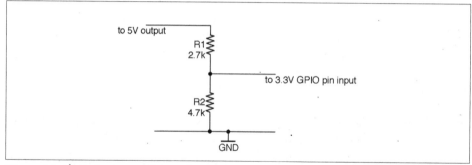

Figure 10-16. Reducing a 5V Signal to 3V

Discussion

Occasionally you will find a 5V device that specifies that the logic level for an input should be above 3.3V. For example, WS2812 LED ICs (Recipe 14.8) theoretically need an input greater than 4V to be considered as logic high according to their datasheet. In practice, I have always found them to work without level conversion, but if you are designing a product to sell, then you would not take that chance and the kind of level shifter shown in Figure 10-17 should be used.

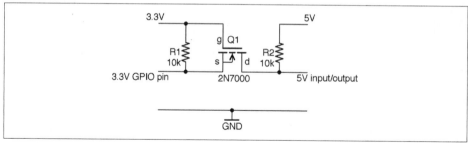

Figure 10-17. Bidirectional Level Shifting with a MOSFET

This circuit will actually convert levels in both directions; that is, 5V outputs will be reduced to 3.3V and 3.3V outputs increased to 5V. It makes use of the fact that MOS-FETs have a "substrate" protection diode that prevents current from flowing from drain to source.

To understand how this circuit works, first consider the case where the 3.3V side is an output and the 5V side an input; that is, the level is being shifted up.

In this case, if the 3.3V GPIO is high, then the gate-source voltage will be 0V, the MOSFET will be off, and R2 will pull-up the 5V input HIGH. If the 3.3V GPIO output is low, the gate-source voltage will be 3.3V, the MOSFET will turn on, and the 5V input will be effectively connected to the LOW signal from the 3.3V side.

If we swap the direction so that the 5V side is now the output and the 3.3V side the input, then when the 5V output is HIGH the MOSFET source and gate will be at 3.3V, the MOSFET will be off, and the 3.3V input pulled up to 3.3V by R1. If the 5V output is LOW the MOSFET's built-in protection diode will conduct, pulling the 3.3V input down to the diode's forward voltage (about 0.6V). This turns the MOSFET on, pulling the 3.3V input fully to ground.

See Also

The simplest level conversion using just two resistors is a voltage divider as described in Recipe 2.6.

For a general background on MOSFETs see Recipe 5.3.

For an interesting discussion of the necessity of level shifting for Raspberry Pi inputs, see *http://tansi.info/rp/interfacing5v.html*.

If you have a lot of signals all requiring level shifting, then it makes sense to use a level-shifting IC or module such as these products from Adafruit: *http://bit.ly/2lLHmuG* (four signals) and *http://bit.ly/2msMgku* (eight signals).

## CHAPTER 11

Switching

11.0 Introduction

Most modern electronics are concerned with switching things. Microcontrollers and any other digital logic device use transistors as switches. Taking this a step further, devices like Arduino and Raspberry Pi can switch external devices, perhaps controlling lighting or power to a heater.

The recipes in this chapter are all concerned with using transistors and other devices to switch. This includes some recipes to extend the switching power of an Arduino and Raspberry Pi.

11.1 Switch More Power than Your Pi or Arduino Can Handle

Problem

You want to allow a pin, such as a microcontroller's GPIO pin, to control more power than it otherwise could.

Solution

Use a transistor in a *common-emitter* arrangement as a "low-side" switch with a resistor to limit the base current. Figure 11-1 shows the schematic for this circuit, that, along with Recipe 11.2, you will use time and time again in your designs.

This type of switching is called "low-side" switching because the transistor acts as a switch between the low voltage of GND and the load.

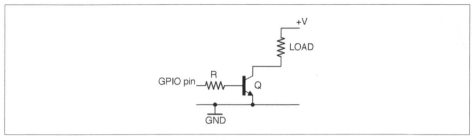

Figure 11-1. Using a Transistor as a Switch

Discussion

The resistor R makes sure that the current drawn from the GPIO pin does not exceed the current limit of the pin (40mA for Arduino and 16mA for Raspberry Pi; see also Recipe 10.7). It also protects the transistor from too much base current flowing. This protective role must be balanced against the limited gain of a bipolar transistor, which may only be 100 or less. So if you are planning to use the transistor to switch a current of 1A, you could reasonably expect a base current of 10mA. This means that for this 1A load, you need to choose a value of R that gives a base current somewhere between the available current of the GPIO and 10mA.

Aiming for a 10mA base current (if you are using a 5V GPIO pin and assume that the base-emitter voltage is a fairly constant 0.6V) you can calculate the value of R as:

$$R = \frac{V}{I} = \frac{5V - 0.6V}{10mA} = 440\Omega \approx 470\Omega$$

Applying the same calculation for a 3.3V GPIO pin you get:

$$R = \frac{V}{I} = \frac{3.3V - 0.6V}{10mA} = 270\Omega$$

The two connections from this circuit to an Arduino or Raspberry Pi are the GND connection and the GPIO pin. The positive voltage supply to the load is entirely separate from the Raspberry Pi or Arduino. This means that you are not restricted to switching the logic level of the Arduino or Pi (3.3V or 5V) but can switch up to the maximum voltage that the transistor is capable of switching.

Having said that, you will often use the positive voltage supply of an Arduino or Raspberry Pi for the convenience of having a single voltage source.

See Also

For a discussion of bipolar transistors, see Recipe 5.1.

GPIO ports and their output logic are described in Recipe 10.7.

You will find an example of switching with a transistor using an Arduino in Recipe 11.6 and for a Raspberry Pi in Recipe 11.7.

For switching using a MOSFET, see Recipe 11.3.

11.2 Switch Power On the High Side

Problem

You want to switch using a BJT, but you need one end of the load to be connected to ground.

Solution

The arrangement of Figure 11-1 is called low-side switching because the switching takes place on the lower voltage side; that is, to GND rather than from the positive supply. This means that whatever is being switched has to have a connection to the positive supply voltage.

One approach to high-side switching is to simply rearrange Figure 11-1 a little as shown in Figure 11-2.

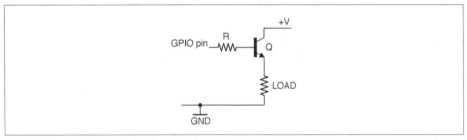

Figure 11-2. High-side Switching with an NPN BJT (Restricted Switching Range)

The circuit of Figure 11-2 cannot switch voltages (+V) greater than the GPIO pin's voltage less half a volt.

A transistor in this arrangement is called an emitter follower (Recipe 16.4) because the emitter will generally be about 0.6V less than the base. This means that we can use this arrangement for high-side switching but only so long as the voltage at the GPIO pin is less than +V.

In short, the circuit of Figure 11-2 is only useful for switch loads where one end of the load must be connected to ground and the other does not need a supply voltage that is greater than the control voltage of the GPIO pin.

Discussion

If you need to switch a higher voltage (perhaps 12V) you might be tempted to consider an alternative to Figure 11-2 that uses a PNP transistor rather than an NPN transistor as shown in Figure 11-3.

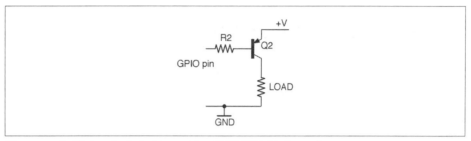

Figure 11-3. High-side Switching with a PNP BJT (Defective)

This circuit is defective because Q2 will turn on with a base current when its base is about 0.6V lower than +V. So, if you were trying to switch a 12V load from 5V logic, the base voltage would always be over 0.6V and enough base current would flow to keep the transistor on whether the GPIO pin was at 5V or GND.

High-side Switching with Tristate Logic

Having said that the schematic of Figure 11-3 is defective, there is a neat trick you can do with a microcontroller to make it work over the full voltage range. The trick relies on being able to change a GPIO pin from being an output to being an input from your program. It works like this:

If the pin is an input, then almost no current will flow into the base of Q2 and so the switch will be off. If you set the GPIO pin to be a low output, then Q2 will turn on.

The only real downside to this trick is that the controlling program will look very confusing to anyone trying to work out what is going on.

The term *tristate logic* simply means that the GPIO pin can be in one of three states: output high, output low, or input (floating).

If you need to be able to switch voltages of V+ that are greater than the 5V of your Arduino or 3.3V of your Raspberry Pi and you don't want to use the tristate logic trick, then you will need to use the arrangement shown in Figure 11-4.

This uses a separate NPN transistor Q1 to control the base current to Q2 giving a full switching range.

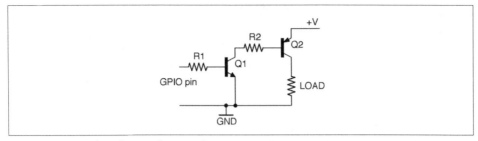

Figure 11-4. High-side Switching with an NPN BJT Driving a PNP BJT

Low-side switching (Recipe 11.1) is the most common and simplest arrangement, and unless you have a good reason such as the need for one end of the load to be connected to ground, you should use low-side switching.

See Also

For a discussion of NPN and PNP bipolar transistors, see Recipe 5.1.

GPIO ports and their output logic are described in Recipe 10.7.

You will find an example of switching with a transistor using an Arduino in Recipe 11.6 and for a Raspberry Pi in Recipe 11.7.

For switching using a MOSFET, see Recipe 11.3.

11.3 Switch Much More Power

Problem

You want to allow a GPIO pin to control more power than it otherwise could, but a BJT isn't enough.

Solution

You can use a MOSFET as an electronic switch. Use the transistor in a *common-source* arrangement. Figure 11-5 shows the schematic for this circuit. Along with Recipe 11.1, you will find yourself using this circuit a lot.

This type of switching is called "low-side" switching because the transistor acts as a switch between the low voltage of GND and the load.

If the GPIO pin is high (3.3V or 5V) and exceeds the gate-threshold voltage of the MOSFET, the MOSFET will turn on, allowing current to flow from +V through the load to GND.

Choosing Logic-level MOSFETs

When using MOSFETs as switches controlled by GPIO pins you should look for devices described as *logic level*. These have a low-threshold voltage (generally 2V or less), whereas MOSFETs not described as logic-level may well have gate-threshold voltages of 4V to 7V.

Logic-level versions of regular MOSFETs often have the same name, but an L after the name. For example, the FQP30N06L is the logic-level version of the FQP30N06.

Low-power MOSFETs like the 2N7000 generally have pretty low gate-threshold voltages anyway and are fine with 3.3V logic without having the L designation.

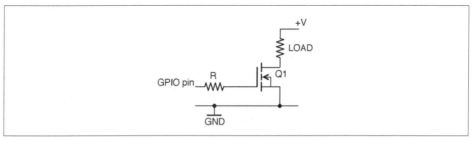

Figure 11-5. Switching with an N-Channel Enhancement-mode MOSFET

Discussion

You may be wondering why the gate resistor R is needed. In reality for most MOS-FETS in most relatively low-current applications, it isn't needed at all and you can just connect the GPIO pin directly to the gate.

However, the MOSFET's gate appears to the GPIO pin as a capacitor that will be charged and discharged as the GPIO pin goes high and low. So, if you are switching the MOSFET at high speed (say PWM at a high frequency) or the MOSFETs' gate capacitance is high (it increases with load current) then the maximum rated output current of the GPIO pin may be exceeded and the transistors inside the GPIO output could overheat. In any case, resistors are cheap and it's good practice to include one.

In Figure 11-5 the gate of the MOSFET is floating. If it is connected to a GPIO set to be an output, then the MOSFET will be in a stable state, but there are several reasons why this might not be the case:

- The program to control the GPIO might not be running yet on your Raspberry Pi, in which case the GPIO will be a floating input.
- The control hardware may be detachable from the Arduino or Raspberry Pi GPIO pin while still retaining its positive supply.

- If the gate is left floating the load is likely to turn itself on and off in response to electrical noise. This is just like the experience of a floating input in an Arduino as described in Recipe 10.10.

To prevent this unwanted behavior, you can just add a pull-down resistor to keep the gate low unless the GPIO pin is actively driving the gate (see Figure 11-6). This resistor should have a resistance of perhaps 10 times that of the gate-protection resistor (R1) to prevent the effective voltage divider (Recipe 2.6) formed by R1 and R2 from reducing the gate voltage by any significant amount.

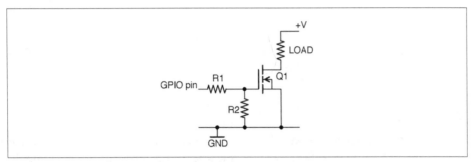

Figure 11-6. Preventing a Floating Gate from Causing Problems

See Also

For a discussion of MOSFETs, see Recipe 5.3.

GPIO ports and their output logic are described in Recipe 10.7.

You will find an example of switching with a MOSFET using an Arduino in Recipe 11.6 and for a Raspberry Pi in Recipe 11.7.

For switching using a BJT, see Recipe 11.1.

11.4 Switch Much More Power on the High Side

Problem

You want to switch using a MOSFET, but you need to share a ground connection between the load being switched and the Arduino or Raspberry Pi doing the switching.

Solution

The problems with high-side switching using a MOSFET are similar to those of Recipe 11.2. However, whereas in Recipe 11.2 it was okay to use an emitter-follower configuration or PNP transistor on its own, the need for a gate-source voltage above the threshold voltage to turn the MOSFET on means that the switchable range is reduced by at least the threshold voltage, making the circuit unsuitable for most applications.

So, for high-side switching there is little alternative to using the circuit of Figure 11-7.

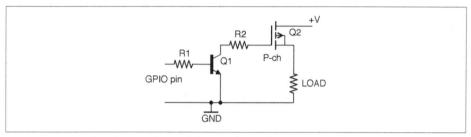

Figure 11-7. High-side Switching with a P-Channel MOSFET and BJT

Discussion

By using a BJT rather than a MOSFET for Q1, there will always be enough leakage current through Q1 to keep the gate of Q2 low when the control pin is left floating.

See Also

For a discussion of MOSFETs, see Recipe 5.3.

GPIO ports and their output logic are described in Recipe 10.7.

You will find an example of switching with a MOSFET using an Arduino in Recipe 11.6 and for a Raspberry Pi in Recipe 11.7.

11.5 Choose Between a BJT and MOSFET

Problem

You can't decide whether to use a BJT or a MOSFET for GPIO-controlled switching.

Solution

Use Table 11-1 to help you decide which NPN BJT or N-Channel MOSFET to use in low-side switching.

Table 11-1. Choosing a transistor

	Technology	Examples
under 100mA	BJT or MOSFET	2N3904 or 2N7000
under 200mA	MOSFET	2N7000

	Technology	Examples
under 500mA	Darlington or MOSFET	MPSA14 or FQP30N06L
under 3A	Darlington or MOSFET	TIP120 or FQP30N06L
under 20A	MOSFET	FQP30N06L

While any of these transistors would technically work, over time you'll find that you return to a few of your favorite transistors.

Discussion

When it comes to switching small loads (say under 100mA) a small BJT (like the 2N3904) and a small MOSFET (like the 2N7000) are pretty much interchangeable. In fact, their pinouts mean that they are in the same transistor package and are pin-compatible.

Above 100mA but below 200mA I would usually select a 2N7000 because its low on-resistance means that it will not get as hot as a BJT.

Between 200mA and 500mA, I would either use a Darlington MPSA14 to save space over a FQP30N06L MOSFET (in a larger package), but only if I could afford the 1.5V or more voltage drop of the MPSA14.

Between 500mA and 3A you are definitely in large-transistor territory in a TO-220 package. If you can afford the voltage drop of a Darlington then the TIP120 is a good choice; otherwise, a FQP30N06L will serve you well.

For powers over 3A use a FQP30N06L with a heatsink.

See Also

See also Recipe 5.5.

2N3904 datasheet: *https://www.sparkfun.com/datasheets/Components/2N3904.pdf*

2N7000 datasheet: *https://www.fairchildsemi.com/datasheets/2N/2N7000.pdf*

MPSA14 datasheet: *http://www.farnell.com/datasheets/43685.pdf*

FQP30N06L datasheet: *https://www.fairchildsemi.com/datasheets/FQ/FQP30N06L.pdf*

TIP120 datasheet: *https://www.fairchildsemi.com/datasheets/TI/TIP122.pdf*

11.6 Switch with Arduino

Problem

You want to use an Arduino to switch a load on and off; that is, more current or a higher voltage than the Arduino GPIO pin can handle.

Solution

If you have a 5V Arduino like the Arduino Uno, then each GPIO pin can switch a maximum of 5V at 40mA. To increase either the voltage or the current, you need to use a transistor. The principles and even the breadboard layouts are much the same whether you use a BJT or a MOSFET. See Recipe 11.5 for the relative merits of each technology.

Figure 11-8 shows the schematic for this.

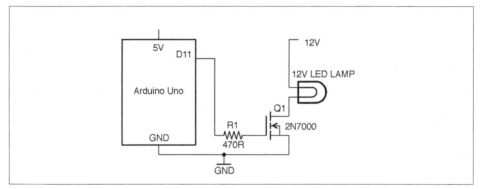

Figure 11-8. Schematic for an Arduino Controlling a 12V Load

NOT for AC

This circuit is for low-voltage DC only. Do *not* try and use it to switch AC. To do so would not work and would also be very dangerous.

Discussion

You can build this circuit on a breadboard to experiment. Figure 11-9 shows the breadboard layout for the project using a 2N7000, which is good for 200mA (2.4W). If you prefer, you could use a 2N3904 or an MPSA14 with exactly the same breadboard layout.

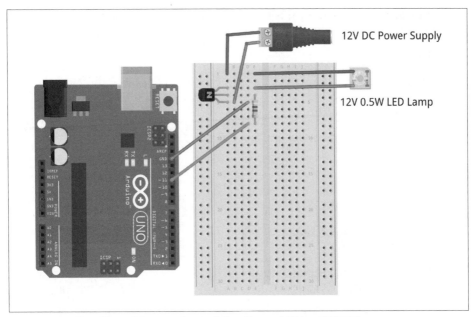

Figure 11-9. Breadboard Layout for Arduino Control Using 2N7000

If you have a higher-power LED lamp, you should use a FQP30N06L as shown in Figure 11-10. Again, you could if you prefer use a TIP120.

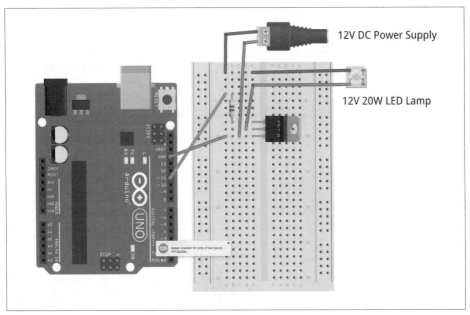

Figure 11-10. Breadboard Layout for Arduino Control Using FQP30N06L

The Arduino test program for this project is listed here and is available in the book downloads (see Recipe 10.2) where the sketch is called *ch_11_on_off*:

```
const int outputPin = 11;

void setup()
{
  pinMode(outputPin, OUTPUT);
  Serial.begin(9600);
  Serial.println("Enter 0 for off and 1 for on");
}

void loop()
{
  if (Serial.available())
  {
    char onOff = Serial.read();
    if (onOff == '1')
    {
      digitalWrite(outputPin, HIGH);
      Serial.println("Output ON.");
    }
    else if (onOff == '0')
    {
      digitalWrite(outputPin, LOW);
      Serial.println("Output OFF.");
    }
  }
}
```

This sketch follows a similar pattern to that of Recipe 10.13. Pin 11 is set to be an output and when a single-command character "1" or "0" is sent from the Serial Monitor (Figure 11-11) `digitalWrite` is used to turn pin 11 on or off.

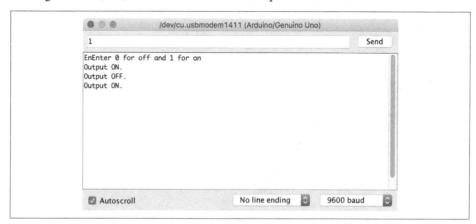

Figure 11-11. Using the Serial Monitor to Turn a Lamp On and Off

You can see the circuit in action in Figure 11-12, where a 12V SLA battery is used to power the lamp.

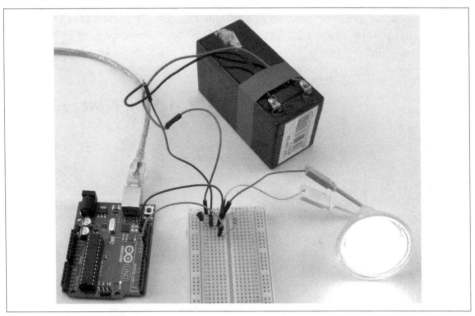

Figure 11-12. Switching a 12V DC Lamp from an Arduino

Although intended as a demonstration circuit, this recipe is actually quite powerful, allowing you to control the lamp from your computer.

See Also

For information on getting started with breadboard, see Recipe 20.1.

For transistor pinouts, see Appendix A.

The basics of an Arduino Uno are described in Recipe 10.1.

To use this same design with Raspberry Pi, see Recipe 11.7.

11.7 Switch with a Raspberry Pi

Problem

You want to use a Raspberry Pi to switch a load on and off; that is, more current or a higher voltage than the GPIO pin can handle.

Solution

The circuits used in Recipe 11.6 will also work just fine if you wire the control connection and GND to a Raspberry Pi rather than an Arduino. Figure 11-13 shows the breadboard layout for using a 2N7000 MOSFET with a Raspberry Pi to switch a 12V LED lamp module.

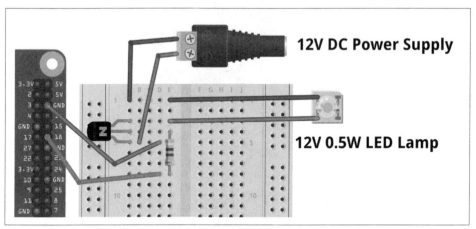

Figure 11-13. Breadboard Layout for Raspberry Pi Control Using 2N7000

To control the lamp, you can use the following program, which is called *ch_11_on_off.py* and can be found with the downloads for the book (see Recipe 10.4):

```python
import RPi.GPIO as GPIO

GPIO.setmode(GPIO.BCM)

led_pin = 18

GPIO.setup(led_pin, GPIO.OUT)


try:
    while True:
        answer = input("1 for on 0 for off: ")
        if answer == 1:
            GPIO.output(led_pin, True)  # LED on
        elif answer == 0:
            GPIO.output(led_pin, False) # LED off
finally:
    print("Cleaning up")
    GPIO.cleanup()
```

When you run the program you should see this in the terminal session:

```
$ sudo python ch_11_on_off.py
1 for on 0 for off: 1
1 for on 0 for off: 0
```

When you enter 1, the lamp should light (pressing 0 will turn it off again).

Discussion

As with Recipe 11.6 you can also use other transistors than the 2N7000 or FQP20N06L MOSFETs.

See Also

For information on getting started with breadboards, see Recipe 20.1.

For transistor pinouts, see Appendix A.

The basics of a Raspbery Pi are described in Recipe 10.3.

To use this same design with an Arduino, see Recipe 11.6.

11.8 Reversible Switching

Problem

You want to be able to switch a load on both the high and low sides because you want to be able to reverse the flow of current through your load, perhaps to control the direction of a DC motor.

Solution

Use transistors in the *half-bridge* configuration shown in Figure 11-14.

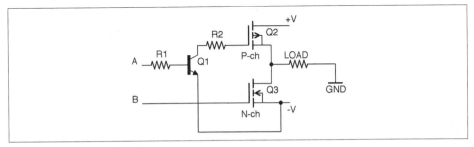

Figure 11-14. A Half-Bridge

This circuit combines elements of Recipe 11.3 and Recipe 11.4 and relies on their being three power lines: +V, –V, and GND in the middle. Thus, the three power lines might be +6V, –6V, and 0V.

The control signals A and B turn on Q2 and Q3, respectively. If A and B are both low with respect to –V no current will flow through the load. If A is high, Q2 will conduct and current will flow through Q2 to the load and then to GND. Conversely, if A is low and B is high, Q3 will conduct and current will flow from GND through the load to –V.

Under no circumstances should A and B both be high, since if both Q2 and Q3 are conducting there will effectively be a short-circuit between +V and –V and a damagingly large current will flow. In fact, if you are connecting A and B to GPIO pins and using software to set them high or low, you should put a small delay in between settings to allow A and B to be high at the same time. For example, to change from A high, B low to A low, B high, you should follow these steps:

1. Set A low
2. Delay
3. Set B high

Discussion

The half-bridge and its relative the full-bridge are often used to control DC motors, where their ability to reverse the direction of the flow of current allows them to reverse the direction in which the motor rotates.

Although it is perfectly possible to build half-bridge circuits from discrete components, it is more common to use an IC. You will meet several of these in Chapter 13.

See Also

For a full-bridge circuit, see Recipe 13.3.

For the FQP27P06 P-Channel MOSFET's datasheet, see *https://www.sparkfun.com/datasheets/Components/General/FQP27P06.pdf.*

11.9 Control a Relay from a GPIO Pin

Problem

You want to switch a relay on and off using an Arduino or Raspberry Pi GPIO pin.

Solution

Use a low-power BJT or MOSFET to switch the power to the relay's coil as shown in Figure 11-15.

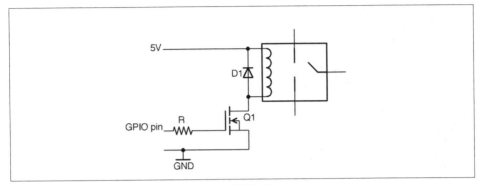

Figure 11-15. Controlling a Relay from a GPIO Pin

A typical relay coil will need about 50mA to activate the relay contacts, which is slightly too much to use an Arduino pin directly and far too much for a Raspberry Pi GPIO pin. A relay with a nominal 5V coil will generally work with a little over 4V if you use a BJT, but supplying the full 5V by way of a MOSFET is generally better. So a good choice for Q1 would be a 2N7000. R is not at all critical, but 1kΩ would be fine.

The diode D1 is called a "freewheeling" or "flyback" diode and provides a discharge path for high voltages generated as the relay coil is switched off, which protects Q1.

Discussion

Building a design like this on a breadboard can be a little tricky, as most common relay packages don't have pins that will fit into the breadboard, so you may have to solder short extension wires to the relay coil or use a prototyping setup like the MonkMakes Protoboard (*https://monkmakes.com/pb*) that has a place to solder a relay. Figure 11-16 shows the breadboard layout for using a relay with an Arduino and Figure 11-17 for a Raspberry Pi.

To test out the relay on an Arduino and Raspberry Pi, you can use the programs described in Recipe 11.6 and Recipe 11.7, respectively.

Although it can be tempting to try to get away with connecting the relay coil directly to a GPIO pin, this may damage the Arduino or Pi and is most certainly operating outside of the specifications for the Arduino, Raspberry Pi, or relay and cannot be guaranteed to work correctly.

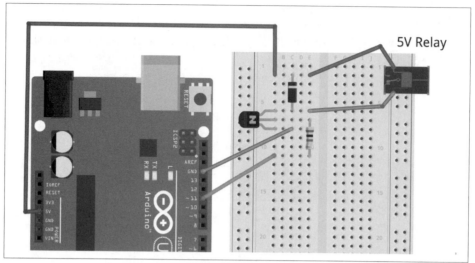

Figure 11-16. Breadboard Layout for Controlling a Relay from an Arduino

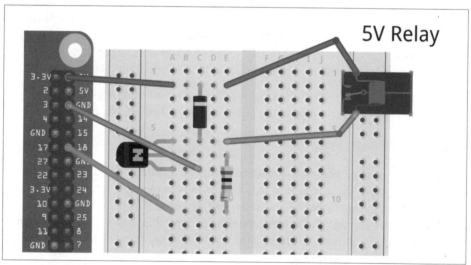

Figure 11-17. Breadboard Layout for Controlling a Relay from a Raspberry Pi

See Also

For information on relays see Recipe 6.4.

11.10 Control a Solid-State Relay from a GPIO Pin

Problem

You want to connect a SSR with an GPIO pin and control the SSR from an Arduino or Raspberry Pi.

Solution

SSRs are generally opto-isolated, which means that controlling them is as simple as using an LED. In fact, it's usually easier because in an SSR a series resistor is generally included. Figure 11-18 shows an SSR module connected to a Raspberry Pi. The SSR's negative input terminal is connected to GND and the positive terminal to a GPIO pin.

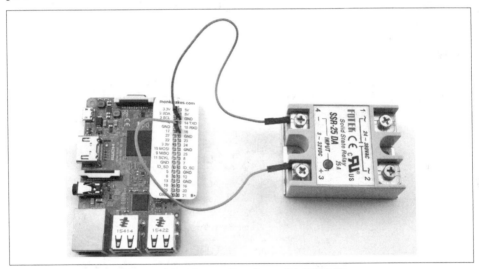

Figure 11-18. Controlling an SSR Using a Raspberry Pi GPIO Pin

Switch AC

Do not use an SSR to control AC unless you are sure you can wire things up safely and you know exactly what you are doing.

Under no circumstances should you work on a live/hot system while you are prototyping, and always use an RCD-protected AC outlet.

See also Recipe 21.12.

Discussion

To test out the relay on an Arduino and Raspberry Pi, you can use the programs described in Recipe 11.6 and Recipe 11.7, respectively.

See Also

A safe and easy way to control AC using an SSR is to use a PowerSwitch Tail (*http://www.powerswitchtail.com/Pages/default.aspx*).

11.11 Connect to Open-Collector Outputs

Problem

You need to know how to use a module (such as a motion detector) or other design that has open-collector outputs.

Solution

As the name suggests, a circuit that has an open-collector output uses an NPN BJT as an output stage (Figure 11-19) with the implication that the emitter is connected to GND, with no internal connections to the collector except to the output.

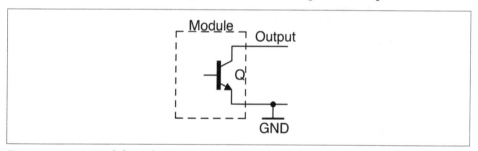

Figure 11-19. A Module with an Open-Collector Output

At first glance this may seem strange, and a common mistake is to treat an open-collector output as a logic output and connect it straight to a GPIO pin. Unless the internal pull-up resistor on the GPIO pin (Recipe 10.7) is enabled, this will not work.

Leaving the collector floating like this has the advantage that by simply using a resistor from the collector to a positive supply, the output voltage of the open collector can be set to match that of the GPIO input you are connecting it to. Figure 11-20 shows how an open-collector output can be used with 3.3V logic (left) and 5V logic (right).

The value of R1 is not critical and needs to lie somewhere in the large range of not exceeding the transistor's collector current limit and not being succeptable to electrical noise. So, anything between 1kΩ and 1MΩ is fine. Common resistor values for this are 1kΩ and 10kΩ.

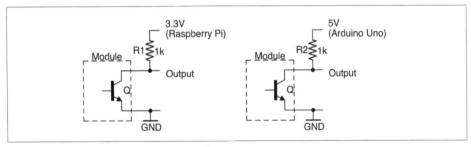

Figure 11-20. Pulling Up an Open-Collector Output

If your GPIO pin has the ability to enable an internal pull-up resistor you can use this instead of an external resistor.

Discussion

If the circuit or module you are using specifies the current rating of the open-collector output, you can use it to directly drive a load (e.g., a relay). The load should be in place of the resistor between the output and a positive voltage supply.

The MOSFET equivalent of an open-collector output is the open-drain output (see Figure 11-21). This can be used in exactly the same way as its BJT equivalent. In fact, it's not uncommon to find device outputs labeled as open collector that actually use a MOSFET.

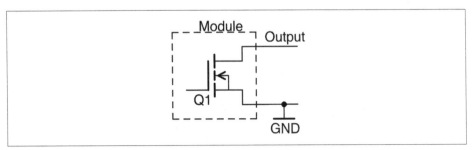

Figure 11-21. An Open-Drain Output

See Also

For more information on BJTs, see Recipe 5.1.

To learn about GPIO pins, see Recipe 10.7.

## CHAPTER 12

Sensors

12.0 Introduction

This chapter looks at sensors that convert some physical measurement such as temperature, light, or physical movement into an analog or digital electronic signal.

Many different sensors are explained, and where appropriate, ways of using them with an Arduino or Raspberry Pi are included.

12.1 Connect a Switch to an Arduino or Raspberry Pi

Problem

You want to be able to convert a mechanical movement to a digital on/off signal that you can use with an Arduino or Raspberry Pi.

Solution

Connect the switch between GND and a GPIO pin set to be a digital input as shown in Figure 12-1 and enable the GPIO pin's internal pull-up resistor.

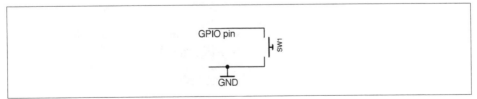

Figure 12-1. Connecting a Switch to a GPIO Pin

Your software will probably also need to "debounce" the signal from the switch.

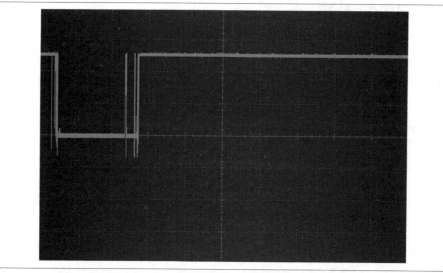

Figure 12-2. Switch Contacts Bouncing

Arduino

Here is an Arduino sketch that writes a message to the Serial Monitor whenever the switch is pressed. You can find this sketch (*ch_12_switch*) with the downloads for the book (see Recipe 7.2):

```
const int inputPin = 12;

void setup()
{
  pinMode(inputPin, INPUT_PULLUP);
```

```
    Serial.begin(9600);
}

void loop()
{
  if (digitalRead(inputPin) == LOW)
  {
    Serial.println("Button Pressed!");
    while (digitalRead(inputPin) == LOW) {};
    delay(10);
  }
}
```

The internal pull-up resistor is turned on by specifying INPUT_PULLUP in the pinMode function.

Inside the loop the inputPin is continually read until it is pressed (becomes LOW). At this point, a message is displayed in the Serial Monitor and the while loop ensures there are no further triggerings until the switch has been released. The debouncing is achieved by the 10-millisecond delay that is triggered after each button press is detected. This gives the switch contacts time to settle before they are tested again.

You can try out the example by poking the pins of a tactile push switch between GND and pin 12 on the Arduino as shown in Figure 12-3.

Figure 12-3. Connecting a Tactile Push Switch to an Arduino

Raspberry Pi

The equivalent code for a Raspberry Pi is listed here (*ch_12_switch.py*):

```
import RPi.GPIO as GPIO
import time

GPIO.setmode(GPIO.BCM)

input_pin = 23

GPIO.setup(input_pin, GPIO.IN, pull_up_down=GPIO.PUD_UP)

try:
try:
    while True:
        if GPIO.input(input_pin) == False:
            print("Button Pressed!")
            while GPIO.input(input_pin) == False:
                time.sleep(0.01)
finally:
    print("Cleaning up")
    GPIO.cleanup()
```

The logic of this program follows exactly the same pattern as for the Arduino.

Having male GPIO pins on a Raspberry Pi makes it a little more difficult to attach a switch. One way to attach the switch is to use a microswitch pushed into a pair of female-to-female jumper wires (Figure 12-4) or use a Squid Button (Figure 12-5).

Figure 12-4. Connecting a Tactile Push Switch to a Raspberry Pi

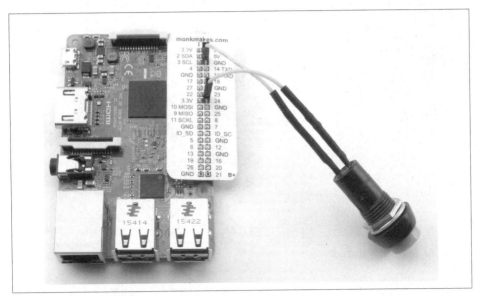

Figure 12-5. Connecting a Squid Button to a Raspberry Pi

If you plan to use a number of push switches, then whether or not you are using a Squid Button, you can use the Squid library to simplify the process of debouncing. Download and install the Squid library from *https://github.com/simonmonk/squid* where you will also find installation instructions. Once installed, the test program can be simplified to the following example (*ch_12_switch_squid.py*):

```python
from button import *
b = Button(23)

while True:
    if b.is_pressed():
        print("Button Pressed!")
```

Discussion

Microswitches (Figure 12-6) are push switches that are not designed to be directly pressed by a person, but rather have a lever attached and are used to sense physical movement such as when a linear actuator has reached its end position or as a safety interlock to make sure a microwave door is closed.

The microswitch is an SPDT device generally with closed, normally open, and common terminals (see Recipe 6.2).

Although switch using an Arduino or Raspberry Pi generally uses a pull-up resistor and switches to ground, you can also switch to the positive supply.

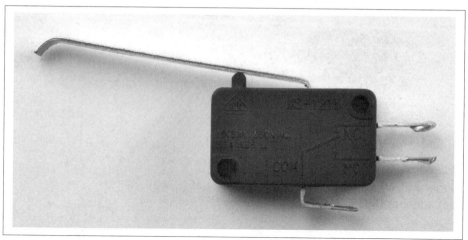

Figure 12-6. A Microswitch

To switch to the positive supply on an Arduino, you have to use an external pull-down resistor to prevent the input from floating because the Arduino does not have internal pull-down resistors on its GPIO pins (pin mode INPUT).

On the other hand, the Raspberry Pi does have internal pull-down resistors that can be enabled for a particular pin using the command:

```
GPIO.setup(input_pin, GPIO.IN, pull_up_down=GPIO.PUD_DOWN)
```

See Also

For information on the Squid Button, see *https://www.monkmakes.com/squid_combo/*.

Digital inputs are also described in Recipe 10.10 and Recipe 10.11.

GPIO pins are described in Recipe 10.7.

12.2 Sense Rotational Position

Problem

You want to use a rotating knob with your Arduino or Raspberry Pi.

Solution

You'll want a type of rotary encoder called a *quadrature encoder*, which behaves like a pair of switches (Figure 12-7). The sequence in which they open and close as the rotary encoder's shaft is turned determines the direction of rotation.

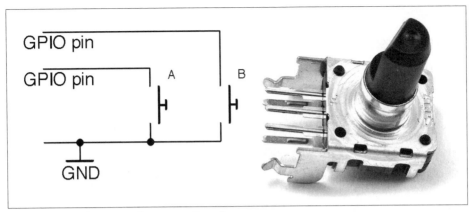

Figure 12-7. Schematic of a Rotary Encoder

A basic rotary encoder will have three pins. One for A, one for B, and a common pin. The rotary encoder shown in Figure 12-7 has two extra pins that are used for a push switch that is activated when the knob is pushed rather than turned.

When using a rotary encoder with a microcontroller or SBC, the common contact is connected to GND and the switch contacts connected to digital inputs with pull-up resistors enabled.

Arduino Software

The following Arduino sketch, which is called *ch_12_quadrature*, can be found with the downloads for the book (Recipe 10.12). The sketch assumes that the two switch pins are connected to pins 6 and 7:

```
const int aPin = 6;
const int bPin = 7;

int x = 0;

void setup()
{
  pinMode(aPin, INPUT_PULLUP);
  pinMode(bPin, INPUT_PULLUP);
  Serial.begin(9600);
}

void loop()
{
  int change = getEncoderTurn();
  if (change != 0)
  {
    x += change;
    Serial.println(x);
  }
```

```
    }

    int getEncoderTurn()
    {
      // return -1, 0, or +1
      static int oldA = 0;
      static int oldB = 0;
      int result = 0;
      int newA = digitalRead(aPin);
      int newB = digitalRead(bPin);
      if (newA != oldA || newB != oldB)
      {
        // something has changed
        if (oldA == 0 && newA == 1)
        {
          result = (oldB * 2 - 1);
        }
        else if (oldB == 0 && newB == 1)
        {
          result = -(oldA * 2 - 1);
        }
      }
      oldA = newA;
      oldB = newB;
      return result;
    }
```

To avoid skipping turns, the function getEncoderTurn must be called as frequently as possible in loop.

getEncoderTurn compares the current state of the A and B switches with their states last time getEncoderTurn was called to infer whether the knob is being turned clockwise or counterclockwise, returning a value of 1 or –1, respectively. If there is no change to the values of A and B then 0 is returned.

Raspberry Pi Software

You can find the Raspberry Pi version of the rotary encoder program in the file *ch_12_quadrature.py* (see Recipe 10.4) and here:

```
import RPi.GPIO as GPIO
import time

GPIO.setmode(GPIO.BCM)

input_A = 18
input_B = 23

GPIO.setup(input_A, GPIO.IN, pull_up_down=GPIO.PUD_UP)
GPIO.setup(input_B, GPIO.IN, pull_up_down=GPIO.PUD_UP)

old_a = 1
```

```
old_b = 1

def get_encoder_turn():
    # return -1, 0, or +1
    global old_a, old_b
    result = 0
    new_a = GPIO.input(input_A)
    new_b = GPIO.input(input_B)
    if new_a != old_a or new_b != old_b :
        if old_a == 0 and new_a == 1 :
            result = (old_b * 2 - 1)
        elif old_b == 0 and new_b == 1 :
            result = -(old_a * 2 - 1)
    old_a, old_b = new_a, new_b
    time.sleep(0.001)
    return result

x = 0

while True:
    change = get_encoder_turn()
    if change != 0 :
        x = x + change
        print(x)
```

The test program counts up as you turn the rotary encoder clockwise, and counts down when you rotate it counterclockwise:

```
pi@raspberrypi ~ $ sudo python rotary_encoder.py
1
2
3
4
5
6
7
8
9
10
9
8
7
6
5
4
```

Discussion

Both the Arduino and Raspberry Pi versions of the software work in just the same way.

Figure 12-8 shows the sequence of pulses that you will get from the two contacts, A and B. You can see that the pattern repeats itself after four steps (hence the name *quadrature* encoder).

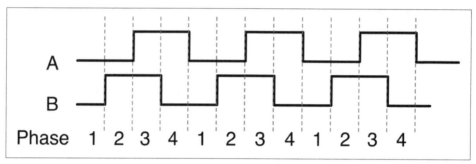

Figure 12-8. How Quadrature Encoders Work

When rotating clockwise (left to right in Figure 12-8), the sequence will be:

Phase	A	B
1	0	0
2	0	1
3	1	1
4	1	0

When rotating in the opposite direction, the sequence of phases will be reversed:

Phase	A	B
4	1	0
3	1	1
2	0	1
1	0	0

The Python program listed previously implements the algorithm for determining the rotation direction in the function get_encoder_turn. The function will return 0 (if there has been no movement), 1 for a rotation clockwise, or -1 for a rotation counterclockwise. It uses two global variables, old_a and old_b, to store the previous states of the switches A and B. By comparing them with the newly read values, it can determine (using a bit of clever logic) which direction the encoder is turning.

The sleep period of 1 millisecond is used to ensure that the next new sample does not occur too soon after the previous sample; otherwise, the transitions can give false readings (see "Contact Bounce" on page 192).

The test program should work reliably no matter how fast you twiddle the knob on the rotary encoder; however, try to avoid doing anything time consuming in the loop, or you may find that turn steps are missed.

See Also

To use a pot to sense rotation, use Recipe 12.3.

12.3 Sense Analog Input from Resistive Sensors

Problem

You want to use a sensor whose resistance varies with some property with an analog input. For example, a photoresistor or thermistor to sense light and temperature, respectively.

Solution

Put the resistive element in a voltage divider with a fixed-value resistor to convert the resistance measurement into a voltage as shown in Figure 12-9.

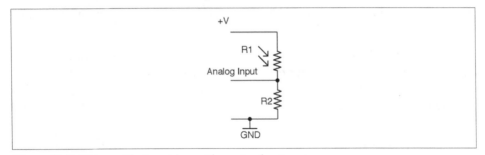

Figure 12-9. Using a Photoresistor with an Analog Input

For Raspberry Pi?

The Raspberry Pi does not have analog inputs, so to use resistive sensors with a Raspberry Pi, you either need to add analog inputs to your Raspberry Pi (see Recipe 12.4) or use the step-response method (see Recipe 12.5).

Discussion

When it comes to choosing the value of a fixed resistor, the aim should be to try and maximize the range of output voltages that will be presented to the analog input. With an Arduino Uno's analog input range of 0 to 5V you ideally want an analog input voltage of 0 to 5V. Because a voltage divider "divides" the input voltage, you are never going to get the full voltage range as an output unless you use a higher input voltage. This introduces the potential for overvoltage at the analog input and so generally it is better to accept a slightly narrowed range rather than risk damaging your board.

Let's take as an example a photoresistor that has a minimum resistance of 1kΩ under bright light and a resistance that rises to 1MΩ in total darkness. Assuming that you are using a 5V Arduino:

$$V_{out} = \frac{R2}{R1 + R2} \cdot V_{in} = 5 \times \frac{R2}{R1 + R2}$$

If we pick a value for the fixed resistor R2 of about 10 times of the minimum value of the photoresistor R1 then R2 would have a value of 10kΩ, so at maximum brightness:

$$V_{out} = 5 \times \frac{R2}{R1 + R2} = 5 \times \frac{10k}{1k + 10k} = 4.55V$$

For a resistance of 10k for the photoresistor, the Vout would be 2.5V and when the photoresistor is in complete darkness (1MΩ), the Vout would fall to 0.05V.

This would give pretty good coverage of the whole range of the photoresistor's readings.

The following Arduino sketch assumes that R1 is a photoresistor with 1kΩ light resistance and R2 is a fixed 10kΩ resistor. The sketch calculates the resistance of R1 and displays it in the Arduino IDE's Serial Monitor every half-second. You can find the sketch here and with the downloads for the book (see Recipe 10.2). It is called *ch_12_r_adc*.

```
const int inputPin = A0;
const float r2 = 1000.0;
const float vin = 5.0;

void setup()
{
  pinMode(inputPin, INPUT);
  Serial.begin(9600);
}

void loop()
{
  int reading = analogRead(inputPin);
  float vout = reading / 204.6;
  float r1 = (r2 * (vin - vout)) / vout;
  Serial.print(r1); Serial.println(" Ohms");
  delay(500);
}
```

See Also

For an introduction to photoresistors, see Recipe 2.8.

For Arduino code to display such analog readings, see Recipe 10.12.

For information on voltage dividers, see Recipe 2.6.

To add analog inputs to a Raspberry Pi, see Recipe 12.4.

12.4 Add Analog Inputs to Raspberry Pi

Problem

You want to measure an analog voltage with a Raspberry Pi.

Solution

The Raspberry Pi does not have analog inputs so connect an ADC IC to the Raspberry Pi.

Use the MCP3008 8-channel ADC chip. This chip actually has eight analog inputs, so you can connect up to eight sensors to one of these and interface to the chip using the Raspberry Pi SPI. Figure 12-10 shows the schematic for using an MCP3008 with a Raspberry Pi.

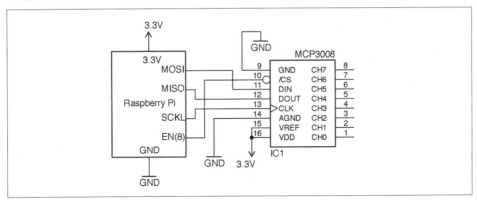

Figure 12-10. Adding Analog Inputs to a Raspberry Pi Using the MCP3008

Note that these analog inputs have a maximum input voltage of 3.3V. You will also need to make sure the SPI of the Raspberry Pi is enabled and the py-spidev Python library (included in new releases of Raspbian) is installed (Recipe 10.16).

Discussion

You can find the Raspberry Pi program to access and display the analog readings from analog channel 0 in the program *ch_12_mcp3008.py* (see Recipe 10.4):

```python
import spidev, time

spi = spidev.SpiDev()
spi.open(0, 0)

def analog_read(channel):
    r = spi.xfer2([1, (8 + channel) << 4, 0])
    adc_out = ((r[1]&3) << 8) + r[2]
    return adc_out

while True:
    reading = analog_read(0)
    voltage = reading * 3.3 / 1024
    print("Reading=%d\tVoltage=%f" % (reading, voltage))
    time.sleep(1)
```

See Also

You can use this recipe with Recipe 12.7, Recipe 12.9, and Recipe 12.10 to read analog sensors with a Raspberry Pi.

You can use resistive sensors with a Raspberry Pi directly, without the need for an ADC IC (see Recipe 12.5).

12.5 Connect Resistive Sensors to the Raspberry Pi without an ADC

Problem

You want to use a resistive sensor with a Raspberry Pi, without the need for an ADC IC.

Solution

See how long it takes for a capacitor to charge the resistive sensor and use this time to calculate the resistance of the sensor.

Figure 12-11 shows the schematic for measuring resistance using nothing more than two resistors and a capacitor.

A Python library to simply read values of resistance can be downloaded from GitHub (*https://github.com/simonmonk/pi_analog*) where you will also find installation instructions and documentation.

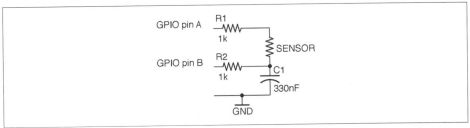

Figure 12-11. Schematic for Measuring Resistance Using the Step-response Method

The following example shows how you can use this library to measure the value of the sensor resistance when the schematic of Figure 12-11 is wired up to a Raspberry Pi with pin A connected to GPIO 18 and pin B to GPIO 23. The program is located in the *examples* folder of the pi_analog library and is called *resistance_meter.py*:

```
from PiAnalog import *
import time

p = PiAnalog()

while True:
    print(p.read_resistance())
    time.sleep(1)
```

To test out the method, try using a few different values of fixed resistance. You can also use different values of C1 and R1 by specifying the value of capacitance in µF and R1 in Ω in the constructor. For example, for high resistance values you might want to use a smaller capacitor (10nF) to speed up the conversion. You could use the following code:

```
from PiAnalog import *
import time

p = PiAnalog(0.01, 1000)

while True:
    print(p.read_resistance())
    time.sleep(1)
```

Discussion

This technique relies on the ability of GPIO pins to switch between being an input and an output while the controlling program is running.

The basic steps for taking a measurement are as follows:

1. Make pin A an input. Make pin B an output and LOW and wait until the capacitor is discharged.
2. Make a note of the time. Make pin B an input and pin A a HIGH output. C1 will now start to charge.
3. When the voltage across C1 reaches about 1.35V it will stop being a LOW input and be measured as HIGH by the GPIO pin connected to B. The time taken for this to happen is a measurement of the resistance of the sensor and R1.

See Also

The Monk Makes Electronics Starter Kit (*https://monkmakes.com/rpi_esk*) for Raspberry Pi has several projects that use this step-response method to measure temperature and light.

12.6 Measure Light Intensity

Problem

You want to measure light intensity with a Raspberry Pi or Arduino.

Solution

If you have an Arduino or board with analog inputs then use Recipe 12.3. If your board does not have analog inputs (say a Raspberry Pi) then use Recipe 12.4 or Recipe 12.5.

Discussion

All the preceding methods will work just fine for determining if the sensor is illuminated more brightly in one reading than in another, but obtaining a measurement in common units of light measurement is a lot trickier. Questions that make getting the light measurement difficult are:

- What frequencies of light are you talking about? All light frequencies, or just the ones the photoresistor is sensitive to?
- Do you want to measure light from a particular direction? If so, the angle of view becomes important.

Even if you answer these questions, you get into problems with the linearity of the photoresistor and will likely need to manually calibrate your meter to account for manufacturing differences in the photoresistors.

In other words, despite its apparrant simplicity making a light meter that will give you a reading in Lux or Watts per square meter is a specialist task.

See Also

For a discussion on light measurement, see *https://en.wikipedia.org/wiki/Lux*.

12.7 Measure Temperature on Arduino or Raspberry Pi

Problem

You want to measure temperature with an Arduino or device with analog inputs.

Solution

Use a thermistor in a voltage-divider arrangement (see Recipe 12.3), calculate the resistance, and then use the Steinhart–Hart equation to calculate the temperature.

Figure 12-12 shows the schematic for connecting a thermistor to an Arduino analog input. You should use an NTC thermistor, which has a specified nominal resistance (its resistance at 25° C) and a value of B (sometimes called beta) that determines its resistance characteristics.

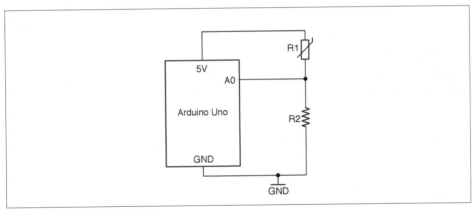

Figure 12-12. Connecting a Thermistor to an Analog Input

You can calculate the resistance of the thermistor R1 using the voltage-divider formula (see Recipe 2.6):

$$V_{out} = \frac{R2}{R1 + R2} \cdot V_{in}$$

This can be rearranged to:

$$R_1 = \frac{R_2\left(V_{in} - V_{out}\right)}{V_{out}}$$

With a 5V Arduino and a fixed 1k resistor for R2, this gives:

$$R_1 = \frac{R_2\left(V_{in} - V_{out}\right)}{V_{out}} = \frac{R_2\left(5 - V_{out}\right)}{V_{out}}$$

The Steinhart–Hart equation states that:

$$\frac{1}{t} = \frac{1}{t_0} + \frac{1}{B}ln\left(\frac{R}{R_0}\right)$$

where:

- t is the temperature in Kelvin (subtract 273.15 for degrees C)
- t0 is 25° C (the standard temperature for the thermistor's base resistance)
- B is a parameter of the thermistor that determines its sensitivity
- R is the resistance of the thermistor at temperature t
- R0 is the resistance of the thermistor at 25° C

If you plug all this math into an Arduino sketch and wire up a thermistor as shown in Figure 12-12, you end up with the following code, which can also be found with the downloads for the book in the sketch *ch_12_thermistor*:

```
const int inputPin = A0;

// r2 is the bottom fixed resistor in the voltage divider
const float r2 = 1000.0;

// thermistor properties
const float B = 3800.0;
const float r0 = 1000.0;

// other constants
const float vin = 5.0;
const float t0k = 273.15;
const float t0 = t0k + 25;

void setup()
{
  pinMode(inputPin, INPUT);
  Serial.begin(9600);
}
```

```
void loop()
{
  int reading = analogRead(inputPin);
  float vout = reading / 204.6;
  float r = (r2 * (vin - vout)) / vout;
  float inv_t = 1.0/t0 + (1.0/B) * log(r/r0);
  float t = (1.0 / inv_t) - t0k;

  Serial.print(t); Serial.println(" deg C");
  delay(500);
}
```

Discussion

Thermistors have become less common for measuring temperature in favor of temperature-sensing ICs. Some ICs are analog devices that produce an output voltage proportional to the temperature like the LM35 and TMP36. Others use a digital interface like the popular DS18B20.

See Also

To measure temperature using a thermistor, the step-response method, and a thermistor, see Recipe 12.8.

You can find an example of using an analog temperature-sensing IC in Recipe 12.10 and a digital device in Recipe 12.11.

12.8 Measure Temperature without an ADC on the Raspberry Pi

Problem

You want to measure temperature using a thermistor, but have a Raspberry Pi, which does not have analog inputs.

Solution

Use a thermistor and the pi-analog library as described in Recipe 12.5. Figure 12-13 shows the schematic for connecting a thermistor to a Raspberry Pi.

The thermistor should be an NTC device. This just means that as the temperature gets hotter the resistance gets lower. You will need to know its nominal resistance (at 25 C) and its value of B.

The pi-analog library has a program for reading the temperature in degrees C that you will find in the *examples* folder for the library called *thermometer.py*. The code for this is listed here:

```
from PiAnalog import *
import time

p = PiAnalog()

while True:
    print(p.read_temp_c(3800, 1000))
    time.sleep(1)
```

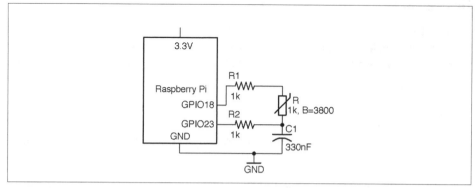

Figure 12-13. Connecting a Thermistor to a Raspberry Pi

The two parameters to `read_temp_c` are the thermistor's value of B and resistance, which should be changed to suit your thermistor.

Discussion

For a device like the Raspberry Pi that does not have analog inputs, an alternative solution is to use an IC with a digital interface like the DS18B20. However, these are considerably more expensive than a thermistor, capacitor, and a couple of resistors.

See Also

To use a thermistor with analog inputs, see Recipe 12.7.

For more information on thermistors, see *https://en.wikipedia.org/wiki/Thermistor*.

You can find an example of using an analog temperature-sensing IC in Recipe 12.10.

12.9 Measure Rotation Using a Potentiometer

Problem

You want to measure rotary position using a pot (variable resistor) with a microcontroller or SBC.

Solution

Option 1

Use the pot as a potential divider, with the slider attached to an analog input as shown in Figure 12-14.

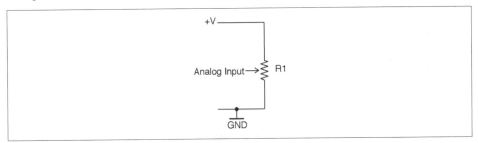

Figure 12-14. Using a Pot to Sense Rotary Position with an Analog Input

The voltage at the slider will vary from 0V to the supply voltage as you rotate the pot's knob. You can use the same test program as Recipe 12.3 to measure the position.

Although a Raspberry Pi does not have analog inputs, you can still use this technique with a Raspberry Pi if you add an ADC chip as described in Recipe 12.4.

Option 2

For devices without an analog input, you can use the step-response method to measure the resistance between one end of the pot and the slider as shown in Figure 12-15.

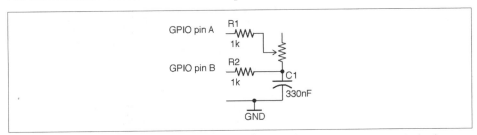

Figure 12-15. Using a Pot to Sense Rotary Position with the Step-response Method

The resistance between the slider and the end of the pot will vary from 0Ω to the maximum resistance of the pot as you rotate the pot's knob. You can use the test program from Recipe 12.5 to measure the resistance of the pot and hence the rotary position.

Discussion

The voltage-divider approach will produce more consistent results than the step-response method.

See Also

For more on pots, see Recipe 2.3.

You can use a quadrature encoder (Recipe 12.2) to measure rotary movement.

For background on voltage dividers, see Recipe 2.6.

12.10 Measure Temperature with an Analog IC

Problem

You want to measure temperature using a linear temperature-sensing IC that produces an output voltage proportional to the temperature.

Solution

Use a temperature-sensing IC such as the TMP36 or LM35. Figure 12-16 shows how you would connect such a device to the analog input of an Arduino. If you want to use such a sensor with a Raspberry Pi, you need to add analog inputs as described in Recipe 12.4. The TMP36 can operate from 3.3V or 5V.

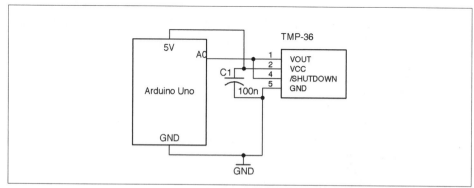

Figure 12-16. Connecting an Analog Temperature-sensing IC to an Arduino Analog Input

Note that only the SMD version of this IC has the extra SHUTDOWN pin. This can be connected to a digital output of, say, an Arduino to reduce the current consumption of the IC to 100nA when the pin is set to LOW.

C1 should be placed as close to the IC as possible.

The temperature of a TMP36 in degrees C (let's call it *t*) is related to the voltage by the following formula:

$$t = 100v - 50$$

where *v* is the output voltage of the TMP36. So, if the output voltage is 0V, the temperature is −50° C; if it is 1V, the temperature is 50° C.

Figure 12-17 shows the breadboard layout for connecting the three-pin through-hole version of the TMP36, connected to an Arduino.

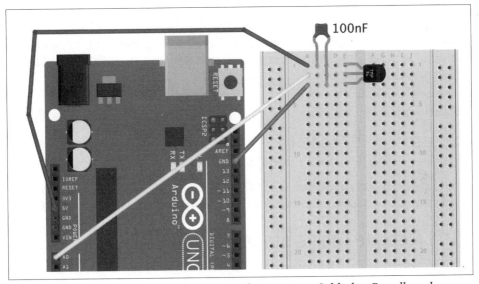

Figure 12-17. Connecting a TMP36 to an Arduino using a Solderless Breadboard

The Arduino sketch *ch_12_tmp36* illustrates the use of this device. You can find this sketch here and with the downloads for the book (see Recipe 10.2):

```
const int inputPin = A0;

const float sensitivity = 0.01; // V/deg C
const float offset = -50.0;     // deg C

void setup()
{
  pinMode(inputPin, INPUT);
```

```
    Serial.begin(9600);
}

void loop()
{
    int reading = analogRead(inputPin);
    float volts = reading / 204.6;
    float degC = (volts / sensitivity) + offset;
    // float degF = degC * 9.0 / 5.0 + 32.0;
    Serial.println(degC);
    delay(500);
}
```

The two constants `sensitivity` and `offset` allow you to use the sketch with other temperature sensors in the TMP36 family of ICs that have different sensitivity and temperature ranges.

Discussion

The TMP36 is not a very accurate device. The datasheet says ±2° C, which is fairly typical in practice and probably with similar accuracy as a thermistor (although the math is a lot easier). For greater accuracy, you should look at a digital sensor like the DS18B20 (see Recipe 12.11).

See Also

For the TMP36 datasheet, see *http://bit.ly/2mbtFsg*.

To measure temperature using a thermistor and an analog input, see Recipe 12.7.

You don't have to have a microcontroller or SBC to make use of a sensor like this. For simple thermostatic applications, you can use a comparator IC (see Recipe 17.10).

12.11 Measure Temperature with a Digital IC

Problem

You want to measure temperature accurately with an Arduino or Raspberry Pi.

Solution

Use a digital temperature measurement IC like the DS18B20, which is factory calibrated to provide an accuracy of ±0.5° C.

This IC uses a 1-wire interface bus that allows up to 255 of the ICs to be connected to just one GPIO port. Figure 12-18 shows a schematic diagram for connecting a DS18B20s to an Arduino. The IC will work just as well with the 3.3V supply of a

Raspberry Pi, although in the case of the Pi, GPIO pin 4 must be used as this pin is earmarked for use by the 1-wire interface.

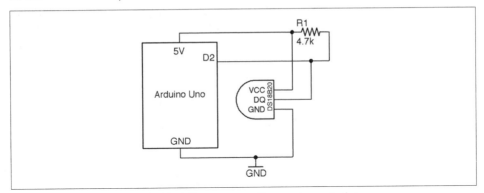

Figure 12-18. Wiring a DS18B20 Temperature Sensor to an Arduino

The 4.7kΩ external pull-up resistor is necessary for the 1-wire bus to operate reliably.

Arduino Software

Having connected the DS18B20 as shown in Figure 12-18, you will need to download the OneWire library (*https://github.com/PaulStoffregen/OneWire*) and the Dallas-Temperature library (*http://bit.ly/2lOYO2j*). In both cases choose the Download ZIP option from the GitHub page and then add the ZIP library to your Arduino IDE from the Sketch→Include Library→Add ZIP Library menu option.

The following sketch illustrates the use of a single DS18B20, reporting the temperature to the Serial Monitor every half-second. You can find the sketch here and with the downloads for the book; it is called *ch_12_ds18b20*:

```
#include <OneWire.h>
#include <DallasTemperature.h>

const int tempPin = 2;

OneWire oneWire(tempPin);
DallasTemperature sensors(&oneWire);

void setup()
{
  Serial.begin(9600);
  sensors.begin();
}

void loop() {
  sensors.requestTemperatures();
  float temp = sensors.getTempCByIndex(0);
```

```
    Serial.println(temp);
  }
```

Because you can attach multiple sensors to a single GPIO pin, the `getTempCByIndex` function takes the number of the sensor that you wish to use (starting at 0). If you have multiple sensors you will have to use trial and error to identify which is which.

Raspberry Pi

Raspbian has support for the 1-wire interface used by the DS18B20 but you have to enable it by editing the file */boot/config.txt* and adding the following line to the end of the file. Then reboot your Raspberry Pi for the change to take effect.

```
dtoverlay=w1-gpio
```

The program for this is called *ch_12_ds18b20.py* and can be found with the downloads for the book (see Recipe 10.4):

```python
import glob, time

base_dir = '/sys/bus/w1/devices/'
device_folder = glob.glob(base_dir + '28*')[0]
device_file = device_folder + '/w1_slave'

def read_temp_raw():
    f = open(device_file, 'r')
    lines = f.readlines()
    f.close()
    return lines


def read_temp():
    lines = read_temp_raw()
    while lines[0].strip()[-3:] != 'YES':
        time.sleep(0.2)
        lines = read_temp_raw()
    equals_pos = lines[1].find('t=')
    if equals_pos != -1:
        temp_string = lines[1][equals_pos+2:]
        temp_c = float(temp_string) / 1000.0
        return temp_c

while True:
    print(read_temp())
    time.sleep(1)
```

As with the Arduino version of this program, the sensor to be read is accessed by position in the line:

```python
device_folder = glob.glob(base_dir + '28*')[0]
```

The interface to the DS18B20 uses a file-like structure. The file interface will always be in the folder */sys/bus/w1/devices/* and the name of the file path will start with 28, but the rest of the file path will be different for each sensor.

The code assumes that there will only be one sensor, and finds the first folder starting with 28. To use multiple sensors, use different index values inside the square brackets.

Within that folder will be a file called *w1_slave*, which is opened and read to find the temperature.

The sensor actually returns strings of text like this:

```
81 01 4b 46 7f ff 0f 10 71 : crc=71 YES
81 01 4b 46 7f ff 0f 10 71 t=24062
```

The code extracts the temperature part of this message. This appears after *t=* and is the temperature in one-thousandths of a degree Celsius.

The read_temp function calculates the temperature in degrees Celsius and returns it.

Discussion

If you want to attach multiple sensors (up to 255) you can wire them up as shown in Figure 12-19. Only the one pull-up resistor is needed, and because the sensor is digital and at a low data transmission rate, you can attach sensors to long leads with no effect on the accuracy of the reading.

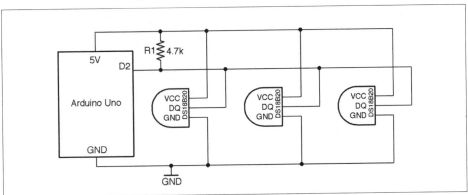

Figure 12-19. Using More than One DS18B20 Temperature Sensor

If you would rather have just two wires running to your DS18B20, you can use the device's special parasitic power feature. In essence, it harvests energy from the data line, storing it temporarily in capacitors inside the IC. In parctical terms, the 3-wire method will allow you to use longer wires to your sensor and will be more reliable, but see the IC's datasheet if you wish to use this mode.

If you plan to deploy your sensor somewhere wet, use encapsulated DS18B20 sensors that are enclosed in a waterproof case that can be found on eBay.

See Also

The DS18B20 datasheet can be found here: *http://bit.ly/2mTPyuu*

To measure temperature with a thermistor, see Recipe 12.7 and Recipe 12.8.

To measure temperature with the TMP36 analog temperature-sensing IC, see Recipe 12.10.

12.12 Measure Humidity

Problem

You want to take humidity readings using an Arduino or Raspberry Pi.

Solution

Use a DHT11 humidity and temperature sensor, which has a serial output that does not really conform to any bus standard like 1-wire, I2C, or SPI. However, libraries for both Arduino and Raspberry Pi are available for it.

Figure 12-20 shows how it should be connected to an Arduino. When used with a Raspberry Pi, the pin labeled VDD (voltage drain drain) should be connected to 3.3V and the DATA pin to GPIO 4.

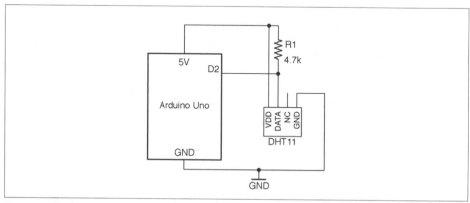

Figure 12-20. Connecting a DHT11 to an Arduino Uno

When connecting to a Raspberry Pi, use the 3.3V power connector *not* the 5V one and connect the data pin of the DHT11 to GPIO4.

Arduino Software

The Arduino sketch can be found with the downloads for the book (Recipe 10.2) and is called *ch_12_dht11*. The sketch requires the Arduino SimpleDHT library to be installed. You can do this from the Arduino Library Manager by using the option Sketch→Include Library→Manage Libraries, searching for SimpleDHT, and then clicking the Install button next to the library.

```
#include <SimpleDHT.h>

const int pinDHT11 = 2;
SimpleDHT11 dht11;

void setup() {
  Serial.begin(9600);
}

void loop() {
  byte temp;
  byte humidity;
  dht11.read(pinDHT11, &temp, &humidity, NULL);

  Serial.print(temp); Serial.print(" C, ");
  Serial.print(humidity); Serial.println(" %");

  delay(1000);
}
```

The call to dht11.read is a little unusual because it passes the variables temp and humidity with an & in front of them to indicate that they can be changed within the read function. In this way, after the read function has finished executing, temp and humidity will have updated values.

Raspberry Pi Software

The Raspberry Pi code for this sensor requires an Adafruit library to be installed using the following commands:

```
$ git clone https://github.com/adafruit/Adafruit_Python_DHT.git
$ cd Adafruit_Python_DHT
$ sudo python setup.py install
```

The program itself can be found with the downloads for the book (Recipe 10.4) and is called *ch_12_dht11.py*:

```
import time, Adafruit_DHT

sensor_pin = 4
sensor_type = Adafruit_DHT.DHT11

while True:
    humidity, temp = Adafruit_DHT.read_retry(sensor_type, sensor_pin)
```

```
print(str(temp) + " C " + str(humidity) + " %")
time.sleep(1)
```

Discussion

The DHT11 is the cheapest and most common series of similar sensors with different accuracies. For better quality results use the DHT22.

See Also

Other recipes for measuring temperature are Recipe 12.7, Recipe 12.8, Recipe 12.10, and Recipe 12.11.

12.13 Measure Distance

Problem

You want to measure the distance from an object to a sensor.

Solution

For distances between 4 inches (10cm) and 6 feet (2m) use an HC-SR04 ultrasonic rangefinder module.

Figure 12-21 shows how the rangefinder is connected to an Arduino. Interfacing to it requires one pin of the Arduino to act as a digital output (connected to TRIG) and one as a digital input (connected to ECHO).

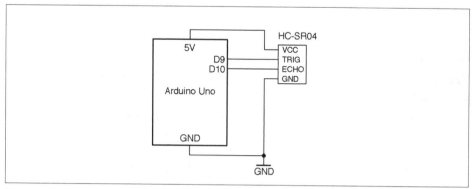

Figure 12-21. Connecting an HC-SR04 Rangefinder to an Arduino

The TRIG pin causes the HC-SR04 to emit a pulse of ultrasound at 40kHz and the ECHO pin goes high when the reflected sound wave is received. The time taken for this to happen is an indication of the distance from the sensor to the object.

To connect the rangefinder to a Raspberry Pi, the output voltage of the ECHO pin of the rangefinder needs to be reduced to 3.3V. This can be done with a simple level converter using a pair of resistors as a voltage divider. The schematic for this is shown in Figure 12-22.

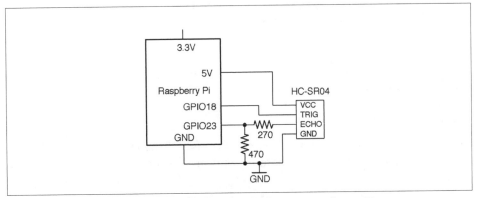

Figure 12-22. Connecting an HC-SR04 Rangefinder to a Raspberry Pi

For more information on level conversion, see Recipe 10.17.

Arduino Software

An Arduino sketch (*ch_12_rangefinder*) for rangefinding can be found with the downloads for the book (see Recipe 10.2):

```
const int trigPin = 9;
const int echoPin = 10;

void setup()
{
  pinMode(trigPin, OUTPUT);
  pinMode(echoPin, INPUT);
  Serial.begin(9600);
}

void loop()
{
  float cm = takeSounding();
  Serial.print(int(cm));
  Serial.print(" cm ");
  int inches = int(cm / 2.5);
  Serial.print(inches);
  Serial.println(" inches");
  delay(500);
}


float takeSounding()
```

```
{
    digitalWrite(trigPin, HIGH);
    delayMicroseconds(10); // 10us trigger pulse
    digitalWrite(trigPin, LOW);
    delayMicroseconds(200); // ingore echos while sending 200us
    long duration = pulseIn(echoPin, HIGH, 100000) + 200;
    float distance = duration / 29.0 / 2.0;
    return distance;
}
```

All the action takes place in the takeSounding function. First, a 10μs pulse is sent to the TRIG pin of the rangefinder resulting in a burst of eight cycles of 40kHz ultrasound. The 200μs delay allows time for this burst to complete before measuring the time taken until an echo is received.

The distance is calculated using the speed of sound (29cm/μs.) The value is divided by 2 because the time taken is for both the outward and return journey of the sound signal.

Raspberry Pi Software

The Raspberry Pi program can be found in the file *ch_12_rangefinder.py* with the downloads for the book (see Recipe 10.4):

```
import RPi.GPIO as GPIO
import time

trigger_pin = 18
echo_pin = 23 # USE LEVEL CONVERTER 5V->3.3V

GPIO.setmode(GPIO.BCM)
GPIO.setup(trigger_pin, GPIO.OUT)
GPIO.setup(echo_pin, GPIO.IN)

def time_to_echo(timeout):
    t0 = time.time()
    while GPIO.input(echo_pin) == False and time.time() < (t0 + timeout):
        pass
    t0 = time.time()
    while GPIO.input(echo_pin) == True and time.time() < (t0 + timeout):
        pass
    return time.time() - t0

def get_distance():
    GPIO.output(trigger_pin, True)
    time.sleep(0.00001)      # 10us
    GPIO.output(trigger_pin, False)
    time.sleep(0.0002)       # 200us
    pulse_len = time_to_echo(1)
    distance_cm = pulse_len / 0.000058
    distance_in = distance_cm / 2.5
```

```
    return (distance_cm, distance_in)



while True:
    print("cm=%f\tinches=%f" % get_distance())
    time.sleep(1)
```

Discussion

The HC-SR04 rangefinders are not particularly accurate, especially when used with a Raspberry Pi, where the operating system will sometimes increase the timing readings for the echo.

Another unwanted influence on the readings comes from variations in the speed of sound. This changes a little both with temperature and humidity.

See Also

The datasheet for the HC-SR04 module can be found here: *http://bit.ly/2mTOPtn*.

## CHAPTER 13

Motors

13.0 Introduction

This chapter looks at various types of motors and how both their speed and direction can be controlled. This involves the electronics of controlling fairly high load currents as well as the software needed. This chapter will cover examples for Arduino and Raspberry Pi.

When people talk of motors, what tends to spring to mind are the small DC motors that you might find in a battery-powered toy car. DC motors are common and often combined with a gearbox to reduce their high speed of rotation into a single unit called a gearmotor.

Stepper motors operate on a different principle and are found in printers of all sorts including 3D printers as they can be advanced in small steps (typically 1/200 of a revolution or more).

Servomotors fit a different niche, allowing accurate positioning of their "arm" in a restricted range of angles (about 180°). Servomotors are often found controlling the steering of a remote-control car or the control surfaces of a remote-control plane or helicopter.

All of these types of motors are available in various sizes to meet the power requirements of different applications. Whatever the size, controlling such motors follows the same basic principles.

13.1 Switch DC Motors On and Off

Problem

You want to control a DC motor using a small control voltage—typically from a GPIO pin.

Solution

Use a transistor switch, but add a snubbing diode as shown in Figure 13-1. This is based on Recipe 11.1, which is fine for a small DC motor. For higher currents use a power MOSFET as described in Recipe 11.3.

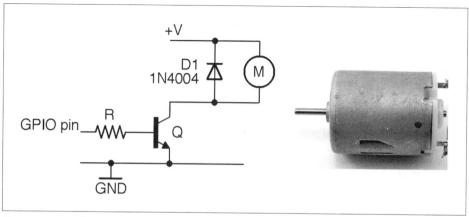

Figure 13-1. Schematic for Switching a DC Motor Alongside a 6V DC Motor

Discussion

The diode D1 prevents the transistor from being damaged. A "snubbing" diode like this should always be used to protect a switching transistor from excessive reverse voltage that occurs when the inductive load of the motor releases stored energy as a pulse of voltage in the opposite polarity to that in which it was driven. This reverse voltage is called "back-emf."

It is not usually necessary to use a snubbing diode with a power MOSFET, because these devices have built-in diodes to prevent damage from electrostatic discharge, which will also protect against voltage spikes. They are (at least for relatively modest motors) sufficient to protect the transistor. However, adding a diode as you would with a BJT is the safest thing to do.

If you want to wire up the circuit of Figure 13-1, you can use the test programs in Recipe 11.6 and Recipe 11.7 to control a motor from an Arduino or Raspberry Pi.

See Also

Many of the techniques described in Chapter 11 can be used to switch a motor.

To control the speed of a motor, see Recipe 13.2.

13.2 Measure the Speed of a DC Motor

Problem

You want to measure the speed of a DC motor.

Solution

Using a transistor switch and PWM to adjust the speed of the motor, measure the motor's speed using an optical detector and slotted wheel attached to the shaft of the motor.

You can use the schematic in Figure 13-2 and pulse the GPIO control pin with PWM as described in Recipe 10.13 and Recipe 10.14.

Discussion

Using PWM to control a motor actually controls the power supplied to the motor rather than its speed, but assuming the motor's load remains constant, these amount to the same thing.

Measuring the motor speed requires a sensor that does not interfere with the rotation of the motor. One way to do this is to use an optical sensor like the prebuilt model shown in Figure 13-2 along with a laser-cut disk that has slots in it. As the disk rotates it alternately blocks and allows light to pass, resulting in a series of pulses that can be used to determine both the speed of the motor and the number of rotations it has made.

The optical sensor includes a built-in comparator and has an "open collector" (Recipe 11.11) output that requires a 1kΩ pull-up resistor to 5V. These sensors are readily available from eBay for a few dollars.

The motor-control circuit uses a MOSFET in the circuit described in Recipe 11.3.

The sketch *ch_13_motor_speed_feedback* (see Recipe 10.2) reports the speed of the motor in RPM (revolutions per minute) once every second in the Serial Monitor. It also allows you to send a PWM value of 0 to 255 to control the speed of the motor.

Figure 13-2. Measuring Motor Speed

Make sure that the Line endings drop-down list of the Serial Monitor is set to "No line ending."

```
const int outputPin = 11;
const int sensePin = 2;

const int slotsPerRev = 20;
const long updatePeriod = 1000L; // ms

long lastUpdateTime = 0;
long pulseCount = 0;

float rpm = 0;

void setup()
{
  pinMode(outputPin, OUTPUT);
  Serial.begin(9600);
  Serial.println("Enter speed 0 to 255");
  attachInterrupt(digitalPinToInterrupt(sensePin), incPulseCount, RISING);
}

void loop()
{
  if (Serial.available())
  {
    int setSpeed = Serial.parseInt();
    analogWrite(outputPin, setSpeed);
  }
  updateRPM();
}

void incPulseCount()
{
```

```
    pulseCount ++;
  }

  void updateRPM()
  {
    long now = millis();
    if (now > lastUpdateTime + updatePeriod)
    {
      lastUpdateTime = now;
      rpm = float(pulseCount) * 60000.0 / (20.0 * updatePeriod);
      pulseCount = 0;
      Serial.println(rpm);
    }
  }
```

An interrupt is used so that every time the sensePin goes from LOW to HIGH the function incPulseCount is called that increments the value of pulseCount.

The function updateRPM will use pulseCount to calculate the RPM once per second before resetting pulseCount to 0.

See Also

To just turn a motor on and off, see Recipe 13.1, and to control the direction of a DC motor, see Recipe 13.3.

13.3 Control the Direction of a DC Motor

Problem

You want to be able to control the direction in which a DC motor turns.

Solution

Use an H-bridge circuit that consists of two push-pull drivers. To keep the number of components low, this is normally accomplished using an H-bridge IC. Figure 13-3 shows the schematic for using the popular L293D motor controller with an Arduino to control two DC motors independently.

The L293D has separate voltage supplies for the ICs logic (VCC1) and the motor (VCC2). This allows the motor to operate at a different voltage than the logic and also reduces the problem of electrical noise caused by the motor disrupting the logic.

Speed control of the motors is accomplished by using a PWM signal to drive the 1,2EN and 3,4EN pins that enable the push-pull drivers in pairs. The direction of one motor is controlled by the L293D pins 1A and 2A and the second motor by 3A and 4A. Table 13-1 shows the four possible states of motor drive for the control pins 1A and 2A.

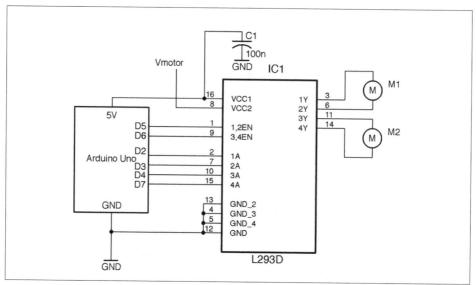

Figure 13-3. Using an L293D IC to Control Two Motors

Table 13-1. Motor Control Pin Logic

1A	2A	Motor M1
LOW	LOW	Off
LOW	HIGH	Clockwise
HIGH	LOW	Counter-clockwise
HIGH	HIGH	Fast-stop

Arduino software

To control the motors, three Arduino pins are needed for each motor. One to control the speed and two for the A and B control pins for the direction that are connected to the 1A and 2A control pins for one motor and the 3A and 4A pins of the L293 for the other.

```
const int motor1SpeedPin = 5;
const int motor2SpeedPin = 6;

const int motor1DirAPin = 2;
const int motor1DirBPin = 3;
const int motor2DirAPin = 4;
const int motor2DirBPin = 7;

void setup()
{
  pinMode(motor1SpeedPin, OUTPUT);
  pinMode(motor2SpeedPin, OUTPUT);
  pinMode(motor1DirAPin, OUTPUT);
```

```
    pinMode(motor1DirBPin, OUTPUT);
    pinMode(motor2DirAPin, OUTPUT);
    pinMode(motor2DirBPin, OUTPUT);
    Serial.begin(9600);
    // M1 full speed clockwise
    analogWrite(motor1SpeedPin, 255);
    digitalWrite(motor1DirAPin, LOW);
    digitalWrite(motor1DirBPin, HIGH);
    // M2 half speed counter-clockwise
    analogWrite(motor2SpeedPin, 127);
    digitalWrite(motor2DirAPin, HIGH);
    digitalWrite(motor2DirBPin, LOW);
}

void loop()
{
}
```

The code sets all the control pins to be outputs and then sets one motor to go at full speed in one direction and the other at half speed in the other. Try experimenting with this code to make the motors behave differently.

Raspberry Pi software

The 3.3V digital outputs of a Raspberry Pi will work just fine connected to the L293D, but the logic supply to the L293D must be 4.5V or more, so the 5V pin of the Raspberry Pi GPIO connector should be used. Table 13-2 shows the connections between Raspberry Pi and L293D.

Table 13-2. Connecting an L293D IC to a Raspberry Pi

Raspberry Pi GPIO Pin	L293D Pin Name	L293 Pin Number(s)	Function
5V	VCC1	16	Logic supply
GND	GND	12	Ground
GPIO18	1,2EN	1	M1 speed
GPIO23	3,4EN	9	M2 speed
GPIO24	1A	2	M1 direction A
GPIO17	2A	7	M1 direction B
GPIO27	3A	10	M2 direction A
GPIO22	4A	15	M2 direction B

The following Python programs (*ch_13_l293d.py*) will set M1 to operate at full speed in one direction and M2 at half-speed in the other direction:

```
import RPi.GPIO as GPIO

GPIO.setmode(GPIO.BCM)
```

```
# Define pins
motor_1_speed_pin = 18
motor_2_speed_pin = 23
motor_1_dir_A_pin = 24
motor_1_dir_B_pin = 17
motor_2_dir_A_pin = 27
motor_2_dir_B_pin = 22

# Set pin modes
GPIO.setup(motor_1_speed_pin, GPIO.OUT)
GPIO.setup(motor_2_speed_pin, GPIO.OUT)
GPIO.setup(motor_1_dir_A_pin, GPIO.OUT)
GPIO.setup(motor_1_dir_B_pin, GPIO.OUT)
GPIO.setup(motor_2_dir_A_pin, GPIO.OUT)
GPIO.setup(motor_2_dir_B_pin, GPIO.OUT)

# Start PWM
motor_1_pwm = GPIO.PWM(motor_1_speed_pin, 500)
motor_1_pwm.start(0)
motor_2_pwm = GPIO.PWM(motor_2_speed_pin, 500)
motor_2_pwm.start(0)

# Set one motor to full speed
motor_1_pwm.ChangeDutyCycle(100)
GPIO.output(motor_1_dir_A_pin, False)
GPIO.output(motor_1_dir_B_pin, True)

# Second motor to half speed
motor_2_pwm.ChangeDutyCycle(50)
GPIO.output(motor_2_dir_A_pin, True)
GPIO.output(motor_2_dir_B_pin, False)

input("Enter '0' to stop ")
print("Cleaning up")
GPIO.cleanup()
```

The code follows the same pattern as its Arduino counterpart. First, the pins are set to be outputs and then two PWM channels are defined to control the motor speeds. Finally, one motor is set to full speed in one direction and the other to half-speed in the other direction.

The function GPIO.cleanup() sets all the pins to be inputs just before the program exits to stop both motors.

Discussion

Figure 13-4 shows a "schematic" for what is going on with an H-bridge driver. Don't try and build this schematic; if you do, prepare for the possibility of destroying some of the transistors if Q1 and Q2 or Q3 and Q4 are turned on at the same time. Also, this circuit will only stand a chance of working if the motor voltage is almost the same as the logic voltage of the control pins (see Recipe 11.4).

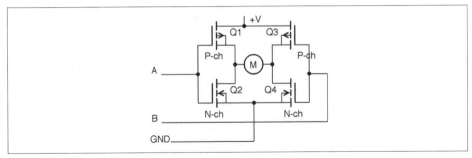

Figure 13-4. A Crude H-Bridge Driver

The idea of the circuit is that the control pin A makes Q1 conduct if it is LOW and Q2 conduct if it is HIGH. Similarly control pin B makes Q3 conduct if it is low and Q4 conduct if pin B is HIGH.

By setting A and B, you can control the direction of flow of current through the motor according to Table 13-1.

See Also

For more information on push-pull (half-bridge) drivers, see Recipe 11.8.

13.4 Setting Motors to Precise Positions

Problem

You want to move a motor to a specific position from an Arduino or Raspberry Pi.

Solution

A servomotor offers the solution to this problem. Wire the servomotor up as shown in Figure 13-5.

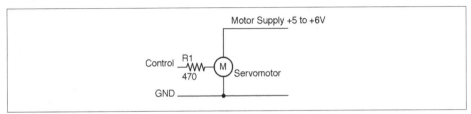

Figure 13-5. Connecting a Servomotor to a GPIO Pin

Generally servomotors will use a separate power supply from the supply used by the Arduino or Raspberry Pi, as the large load current as the motor starts may well drop the supply voltage enough to reset the controlling device. However, for a small lightly loaded servomotor, you may get away with using the same supply for both.

The resistor R1 is there to protect the GPIO pin, but is not strictly necessary as most servomotors draw very little current from the control pin. But if you do not have the datasheet for the servo you are using then R1 is a sensible precaution.

Figure 13-6 shows a small 9g hobby servo. The connector to the servo is fairly standardized, but you should also check the servomotor's datasheet.

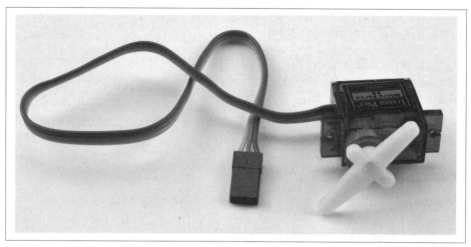

Figure 13-6. A 9g Servomotor

The red lead is the positive supply to the motor, the brown lead the ground connection, and the orange lead the control signal.

The control signal will normally be fine with a 3.3V logic, but if the datasheet for the servomotor indicates that this needs to be higher, use a level converter (Recipe 10.17).

Arduino software

The following example code (*ch_13_servo*) in the downloads for the book (see Recipe 10.2) assumes that the control pin of a servo is connected to pin 9 of an Arduino Uno.

When you open the Arduino Serial Monitor you will be prompted to enter an angle for the servoarm:

```
#include <Servo.h>

const int servoPin = 9;

Servo servo;
```

```
void setup() {
void setup() {
  servo.attach(servoPin);
  servo.write(90);
  Serial.begin(9600);
  Serial.println("Angle in degrees");
}

void loop() {
  if (Serial.available()) {
    int angle = Serial.parseInt();
    servo.write(angle);
  }
}
```

Make sure that you have Line endings drop-down on the Serial Monitor set to "No line ending."

The `servo.write` method of the servo library sets the servo arm to an angle in degrees between 0 and 180.

Raspberry Pi software

The Raspberry Pi equivalent of the Arduino sketch is *ch_13_servo.py* in the book downloads (see Recipe 10.4):

```
import RPi.GPIO as GPIO
import time

servo_pin = 18

# Tweak these values to get full range of servo movement
deg_0_pulse = 0.5   # ms
deg_180_pulse = 2.5 # ms
f = 50.0    #50Hz = 20ms between pulses

# Do some calculations on the pulse width parameters
period = 1000 / f # 20ms
k = 100 / period        # duty 0..100 over 20ms
deg_0_duty = deg_0_pulse * k
pulse_range = deg_180_pulse - deg_0_pulse
duty_range = pulse_range * k

GPIO.setmode(GPIO.BCM)
GPIO.setup(servo_pin, GPIO.OUT)
pwm = GPIO.PWM(servo_pin, f)
pwm.start(0)

def set_angle(angle):
    duty = deg_0_duty + (angle / 180.0) * duty_range
    pwm.ChangeDutyCycle(duty)
```

```
try:
    while True:
        angle = input("Angle (0 to 180): ")
        set_angle(angle)
finally:
    print("Cleaning up")
    GPIO.cleanup()
```

The set_angle function makes use of the variables deg_0_duty and duty_range, which are calculated once at the start of the program to calculate the duty cycle that will generate a pulse of the right length for the specified angle.

Servomotors rarely have exactly the same range of movement, so this program and its Arduino equivalent are a great way to find the range of angles that your servomotor can reach.

Discussion

Figure 13-7 shows how the pulses arriving at a servomotor's control pin alter the angle of the servomotor's arm.

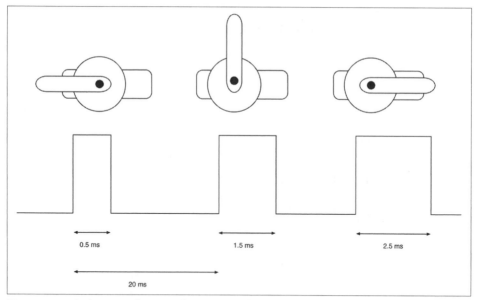

Figure 13-7. Pulse-Length Controls Position of Arm on a Servomotor

The servomotor expects a pulse every 20ms to maintain its position. The length of that pulse determines the position of the servomotor's arm, which can normally travel through about 180°. A short pulse of between 0.5 and 1ms will position the arm at one end of its travel. A pulse of 1.5ms will position the servomotor's arm in the center position and a long pulse of up to 2.5ms at its furthest travel.

Connect Lots of Servos

If you have a lot of servomotors to connect to an Arduino or a Raspberry Pi, then it can help to use a servomotor interface board such as the ServoSix board (Figure 13-8).

Figure 13-8. The MonkMakes ServoSix Board

Using the Python GPIO library to generate the pulses can result in some jitter in the movement of the servoarm. This is due to inaccuracies of the pulse timing as Raspbian tries to do several things at once.

This can be improved by using the ServoBlaster service that configures certain GPIO pins exclusively for servo use or the ServoSix library that is built on ServoBlaster, but it just configures the pins for servo use while the controlling program is running.

Another alternative is to devolve the control of the servos entirely to hardware such as the Adafruit 16-channel servoboard, which is then sent control messages from the Raspberry Pi.

See Also

For more information on the ServoBlaster Python library for Raspberry Pi, see *https://github.com/richardghirst/PiBits/tree/master/ServoBlaster*.

You can find the ServoSix library at *https://github.com/simonmonk/servosix* and the ServoSix connection board at *https://www.monkmakes.com/servosix/*.

To use the Adafruit servo interface board, see *https://www.adafruit.com/product/815*.

13.5 Move a Motor a Precise Number of Steps

Problem

You want to move a motor a precise number of degrees or steps from one position to another using an Arduino or Raspberry Pi.

Solution

A bipolar stepper motor can do what you need. Use a dual H-bridge IC such as the L293D used in Recipe 13.3 to drive the two coils of a bipolar stepper motor. Figure 13-9 shows the schematic diagram for this using an Arduino Uno and Figure 13-10 shows the diagram for a Raspberry Pi.

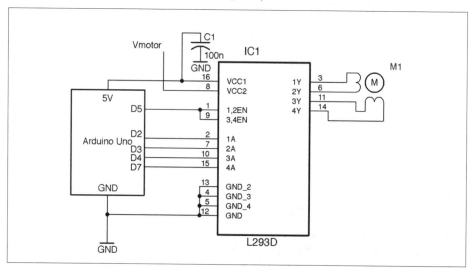

Figure 13-9. Controlling a Bipolar Stepper Motor Using an Arduino Uno

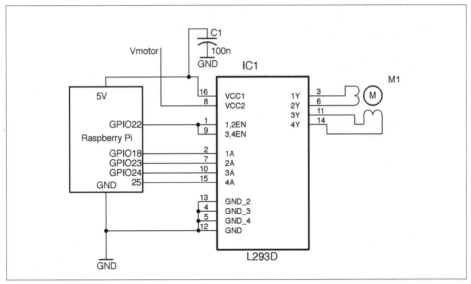

Figure 13-10. Controlling a Bipolar Stepper Motor Using a Raspberry Pi

A push-pull driver is needed for each of the two coils of the stepper motor. The enable pins of the L293D are both tied to a single GPIO pin. Generally the enable pins will be permanently enabled, although they could be used with a high-frequency PWM signal if the stepper motor voltage is below the power-supply voltage.

Arduino software

You can find an Arduino sketch called *ch_13_bi_stepper* in the downloads for the book (Recipe 10.2):

```
#include <Stepper.h>

const int in1Pin = 2;
const int in2Pin = 3;
const int in3Pin = 4;
const int in4Pin = 7;
const int enablePin = 5;

Stepper motor(200, in1Pin, in2Pin, in3Pin, in4Pin);

void setup() {
  pinMode(in1Pin, OUTPUT);
  pinMode(in2Pin, OUTPUT);
  pinMode(in3Pin, OUTPUT);
  pinMode(in4Pin, OUTPUT);
  pinMode(enablePin, OUTPUT);
  digitalWrite(enablePin, HIGH);
  Serial.begin(9600);
```

```
    Serial.println("Command letter followed by number");
    Serial.println("p20 - set the motor speed to 20");
    Serial.println("f100 - forward 100 steps");
    Serial.println("r100 - reverse 100 steps");
    motor.setSpeed(20);
}

void loop() {
  if (Serial.available()) {
    char command = Serial.read();
    int param = Serial.parseInt();
    if (command == 'p') {
      motor.setSpeed(param);
    }
    else if (command == 'f') {
      motor.step(param);
    }
    else if (command == 'r') {
      motor.step(-param);
    }
  }
}
```

The sketch uses the Stepper library that is included with the Arduino IDE. The library requires you to specify the number of steps as the first parameter in the following line:

```
Stepper motor(200, in1Pin, in2Pin, in3Pin, in4Pin);
```

To try out the program, open the Serial Monitor and send commands such as *f100*, where *f* is the direction (forward or reverse) and 100 is the number of steps.

Raspberry Pi software

The Raspberry Pi software for driving a bipolar stepper motor can be found in the file *ch_13_bi_stepper.py*. To download the code for the book see Recipe 10.4.

```
import RPi.GPIO as GPIO
import time

GPIO.setmode(GPIO.BCM)

in_1_pin = 18
in_2_pin = 23
in_3_pin = 24
in_4_pin = 25
en_pin = 22

GPIO.setup(in_1_pin, GPIO.OUT)
GPIO.setup(in_2_pin, GPIO.OUT)
GPIO.setup(in_3_pin, GPIO.OUT)
```

```
GPIO.setup(in_4_pin, GPIO.OUT)
GPIO.setup(en_pin, GPIO.OUT)
GPIO.output(en_pin, True)

period = 0.02

def step_forward(steps, period):
  for i in range(0, steps):
    set_coils(1, 0, 0, 1)
    time.sleep(period)
    set_coils(1, 0, 1, 0)
    time.sleep(period)
    set_coils(0, 1, 1, 0)
    time.sleep(period)
    set_coils(0, 1, 0, 1)
    time.sleep(period)
def step_reverse(steps, period):
  for i in range(0, steps):
    set_coils(0, 1, 0, 1)
    time.sleep(period)
    set_coils(0, 1, 1, 0)
    time.sleep(period)
    set_coils(1, 0, 1, 0)
    time.sleep(period)
    set_coils(1, 0, 0, 1)
    time.sleep(period)

def set_coils(in1, in2, in3, in4):
  GPIO.output(in_1_pin, in1)
  GPIO.output(in_2_pin, in2)
  GPIO.output(in_3_pin, in3)
  GPIO.output(in_4_pin, in4)


try:
    print('Command letter followed by number');
    print('p20 - set the inter-step period to 20ms (control speed)');
    print('f100 - forward 100 steps');
    print('r100 - reverse 100 steps');

    while True:
        command = input('Enter command: ')
        parameter_str = command[1:] # from char 1 to end
        parameter = int(parameter_str)
        if command[0] == 'p':
            period = parameter / 1000.0
        elif command[0] == 'f':
            step_forward(parameter, period)
        elif command[0] == 'r':
            step_reverse(parameter, period)
```

```
finally:
    print('Cleaning up')
    GPIO.cleanup()
```

This code does not use a library, but directly sets the outputs of the push-pull drivers
to activate the coils in the right sequence to move the motor forward or backward for
the specified number of steps.

Using Python 2?

The preceding code is written for Python 3. If you are running the
program as Python 2, then you will need to change the line:

```
command = input('Enter command: ')
```

to:

```
command = raw_input('Enter command: ')
```

The Python programs in this book are designed to work with
Python 2 or Python 3, but the `input`/`raw_input` issue is one source
of incompatibility between the Python versions that is difficult to
code around.

Discussion

Unlike servomotors, stepper motors can rotate continuously—they just do it one step
at a time. Typically a stepper motor will have a few tens to a few hundred steps to one
full rotation. Moving the stepper motor from one step position to the next involves
activating its two coils in a certain pattern.

The L293D lends itself well to experimenting with stepper motors on a breadboard.
Figure 13-11 shows an Arduino wired up from the schematic of Figure 13-9. Note
that the GND connections are connected inside the chip, so not all ground connec‐
tions in the schematic need to be made.

See Also

The stepper motor used to test out the preceding examples is from Adafruit (*https://
www.adafruit.com/products/324*), where you will also find a datasheet for the motor).

To use unipolar stepper motors (usually 5-lead), see Recipe 13.6.

Figure 13-11. Using a Breadboard with an L293D

13.6 Choose a Simpler Stepper Motor

Problem

You want to use a unipolar (5-wire) stepper motor with an Arduino or Raspberry Pi.

Solution

Unipolar stepper motors are a little easier to use than their bipolar relatives discussed in Recipe 13.6. They do not need a push-pull half-bridge driver, but can be controlled using a Darlington array chip like the ULN2803. Figure 13-12 shows the schematic for using this IC with an Arduino and Figure 13-13 for a Raspberry Pi.

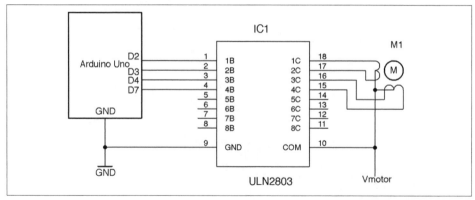

Figure 13-12. Using an ULN2803 to Control a Unipolar Stepper Motor (Arduino)

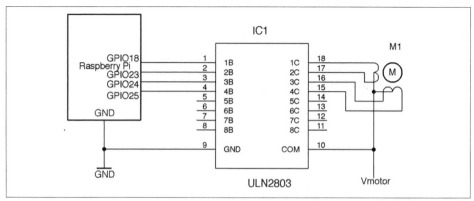

Figure 13-13. Using an ULN2803 to Control a Unipolar Stepper Motor (Raspberry Pi)

The ULN2803 contains eight open-collector Darlington transistors, each capable of sinking about 500mA and so can be used to drive two unipolar stepper motors.

Arduino software

You can find an example Arduino sketch that uses the stepper library in *ch_13_uni_stepper* with the downloads for the book (see Figure 10-2).

```
#include <Stepper.h>

const int in1Pin = 2;
const int in2Pin = 3;
const int in3Pin = 4;
const int in4Pin = 7;

Stepper motor(513, in1Pin, in2Pin, in3Pin, in4Pin);

void setup() {
```

```
    pinMode(in1Pin, OUTPUT);
    pinMode(in2Pin, OUTPUT);
    pinMode(in3Pin, OUTPUT);
    pinMode(in4Pin, OUTPUT);
    Serial.begin(9600);
    Serial.println("Command letter followed by number");
    Serial.println("p20 - set the motor speed to 20");
    Serial.println("f100 - forward 100 steps");
    Serial.println("r100 - reverse 100 steps");
    motor.setSpeed(20);
}

void loop() {
    if (Serial.available()) {
    char command = Serial.read();
    int param = Serial.parseInt();
    if (command == 'p') {
      motor.setSpeed(param);
    }
    else if (command == 'f') {
      motor.step(param);
    }
    else if (command == 'r') {
      motor.step(-param);
    }
  }
}
```

The sketch is pretty much the same as that of Recipe 13.5, but in this case there is no enable pin possible. See the sketch in Recipe 13.5 for a description of the code.

Raspberry Pi software

The Raspberry Pi version of the software can be found in the program *ch_13_uni_stepper.py*. It is identical to that of Recipe 13.5 but with the driver-enable feature removed:

```
import RPi.GPIO as GPIO
import time

GPIO.setmode(GPIO.BCM)

in_1_pin = 18
in_2_pin = 23
in_3_pin = 24
in_4_pin = 25

GPIO.setup(in_1_pin, GPIO.OUT)
GPIO.setup(in_2_pin, GPIO.OUT)
GPIO.setup(in_3_pin, GPIO.OUT)
GPIO.setup(in_4_pin, GPIO.OUT)
```

```python
period = 0.02

def step_forward(steps, period):
  for i in range(0, steps):
    set_coils(1, 0, 0, 1)
    time.sleep(period)
    set_coils(1, 0, 1, 0)
    time.sleep(period)
    set_coils(0, 1, 1, 0)
    time.sleep(period)
    set_coils(0, 1, 0, 1)
    time.sleep(period)
def step_reverse(steps, period):
  for i in range(0, steps):
    set_coils(0, 1, 0, 1)
    time.sleep(period)
    set_coils(0, 1, 1, 0)
    time.sleep(period)
    set_coils(1, 0, 1, 0)
    time.sleep(period)
    set_coils(1, 0, 0, 1)
    time.sleep(period)


def set_coils(in1, in2, in3, in4):
  GPIO.output(in_1_pin, in1)
  GPIO.output(in_2_pin, in2)
  GPIO.output(in_3_pin, in3)
  GPIO.output(in_4_pin, in4)

try:
    print('Command letter followed by number');
    print('p20 - set the inter-step period to 20ms (control speed)');
    print('f100 - forward 100 steps');
    print('r100 - reverse 100 steps');

    while True:
        command = raw_input('Enter command: ')
        parameter_str = command[1:] # from char 1 to end
        parameter = int(parameter_str)
        if command[0] == 'p':
            period = parameter / 1000.0
        elif command[0] == 'f':
            step_forward(parameter, period)
        elif command[0] == 'r':
            step_reverse(parameter, period)

finally:
    print('Cleaning up')
    GPIO.cleanup()
```

Discussion

Unipolar stepper motors are available with built-in reduction gearboxes that make great motors for making small roving robots.

See Also

The unipolar stepper motor that I used to validate these recipes is this Adafruit model: *https://www.adafruit.com/product/858*.

To use bipolar stepper motors, see Recipe 13.6.

The ULN2803 datasheet can be found here: *http://www.ti.com/lit/ds/symlink/uln2803a.pdf*.

## CHAPTER 14

LEDs and Displays

14.0 Introduction

LEDs can be used both for illumination and as indicators. They can also be arranged as 7-segment displays or as tiny pixels in an OLED (organic LED) display.

This chapter contains recipes relating to powering and controlling LEDs as well as for using a display with an Arduino or Raspberry Pi.

14.1 Connect Standard LEDs

Problem

You want to connect a standard low-power LED to a GPIO pin, but you are not sure what value of series resistor to use.

Solution

Back in Recipe 4.4 you saw how an LED needs a series resistor to prevent too much current from flowing. Too much current will shorten the life of an LED, but more importantly, it may damage or destroy the Arduino or Raspberry Pi GPIO pin that is controlling the LED.

Wire up your LED to a GPIO pin as shown in Figure 14-1 and then use Table 14-1 to select a value for the series resistor.

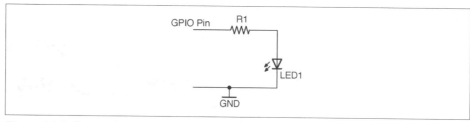

Figure 14-1. Connecting an LED to a GPIO Pin

Table 14-1. LED Series Resistor Values

	Infrared	Red	Orange/Yellow/Green	Blue/White	Violet	UV
Vf	1.2-1.6V	1.6-2V	2-2.2V	2.5-3.7V	2.7-4V	3.1-4.4V
3.3V GPIO 3mA	1kΩ	680Ω	470Ω	270Ω	220Ω	68Ω
3.3V GPIO 16mA	150Ω	120Ω	82Ω	56Ω	39Ω	15Ω
5V GPIO 20mA	220Ω	180Ω	150Ω	150Ω	120Ω	100Ω

Discussion

In practice, nearly all LEDs will illuminate to some degree even from 3.3V with a 1kΩ resistor limiting the current. So, as an even broader rule of thumb, if you don't care about optimizing the brightness of your LED then a 270Ω resistor is fine.

For optimum brightness find the datasheet for your LED and use the forward-voltage Vf and maximum forward current If with the formula in Recipe 4.4 to calculate the optimum value of resistor.

If you connect an LED as shown in Figure 14-1 you can test turning it on and off using an Arduino and Raspberry Pi with Recipe 10.8 and Recipe 10.9.

Figure 14-2. A Raspberry Squid RGB LED

A convenient thing to have around if you want to use LEDs with a Raspberry Pi is a Raspberry Squid (Figure 14-2). This has series resistors for the red, green, and blue LEDs built-in to the leads, so you can just plug it straight onto the GPIO connector of a Raspberry Pi.

Identifying LED Pins

If you take a close look at a standard through-hole LED like the one shown in Figure 14-3 you will see that one lead is longer than the other. The longer lead is the anode should be at the more positive voltage so that the LED is forward-biased and therefore lights.

Looking through the clear plastic enclosure of this LED, you should also be able to see that the connections inside the LED are different shapes, one being much larger than the other.

Finally, if any doubt remains, LEDs generally have one side with a flat edge that indicates the cathode (negative terminal).

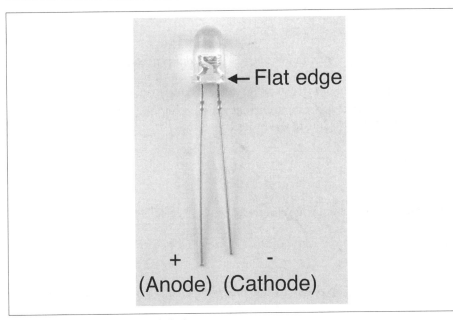

Figure 14-3. A 5mm Through-hole LED

See Also

For more background on LEDs, see Recipe 4.4.

To make your own Raspberry Squid, see *https://github.com/simonmonk/squid*.

14.2 Drive High-Power LEDs

Problem

You want to power a 2–5W LED using a constant current and be able to control the LED's brightness from the PWM GPIO signal on an Arduino or Raspberry Pi.

Solution

Modify Recipe 7.7 to add a control input for the LED driver as shown in Figure 14-4.

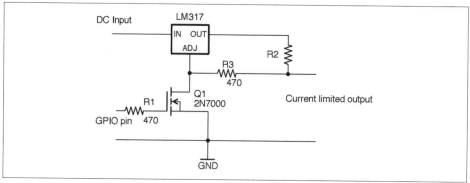

Figure 14-4. Constant-Current LED Control from Arduino or Raspberry Pi

When the GPIO pin is low, Q1 will be off and effectively takes no part in the circuit. The LM317 then acts as a constant-current source where R2 sets the current as discussed in Recipe 7.7 according to the formula:

$$I = \frac{1.2}{R_2}$$

So if you know what current you want to set, you can calculate the value of R2 using:

$$R_2 = \frac{1.2}{I}$$

Since R3 is very low compared to the input impedance of the Adjust pin of the LM317, it can be ignored.

When the GPIO pin is taken above the gate-threshold voltage of Q1, Q1 will turn on, pulling the Adjust pin low and reducing the regulated voltage to 0V.

Discussion

The LM317 will get hot if the DC-input voltage is significantly higher than the forward voltage of the LED. In fact, the power it will turn into heat will be:

$$P = I_f\left(V_{in} - V_f - 1.2\right)$$

where If is the forward voltage of the LED, Vin is the input voltage, and Vf is the forward voltage of the LED.

R2 will generally be a low-value resistor that will produce heat with a power of:

$$P = 1.2 \times I_f$$

As an example, Figure 14-5 shows the circuit of Figure 14-4 supplying 120mA to a high-power LED. For this, R2 is a 10Ω 1/4W resistor. The control pin is connected to a GPIO pin of an Arduino.

Figure 14-5. Controlling a Constant-Current Source from an Arduino

Arduino software

You can turn the LED on and off using the sketch from Recipe 10.8 but on and off will be swapped over because a HIGH at the GPIO pin will turn the LED off.

You can also use PWM to control the LED brightness, but again, the logic is inverted, so you will need to modify the following line in ch_10_analog_output from:

```
int brightness = Serial.parseInt();
```

to be:

```
int brightness = 255 - Serial.parseInt();
```

Raspberry Pi software

You can use the Python program from Recipe 10.9, but as with the equivalent Arduino sketch the on/off logic will be inverted.

For brightness control, you will also need to modify the line:

```
duty = int(duty_s)
```

to be:

```
duty = 100 - int(duty_s)
```

See Also

The LM317 datasheet can be found here: *http://www.ti.com/lit/ds/symlink/lm317.pdf*.

14.3 Power Lots of LEDs

Problem

You want to power a whole load of LEDs at the same time.

Solution

Arrange columns of LEDs in series to match the supply voltage, with a current-limiting resistor for each column. The example in Figure 14-6 shows how you could power 20 LEDs with a forward voltage of 1.7V and a forward current of around 20mA from a 12V power supply.

In this case, five forward-biased LEDs in a column will drop a total of 8.5V, leaving 3.5V to be dropped across the resistor. Using Ohm's Law, this means the resistor should have a value of 175Ω so a standard value of 180Ω would lower the current just slightly but enough.

While it's tempting to reduce the number of resistors you need by putting more LEDs in series, this can cause practical problems of overcurrent if the stated forward voltage of the LEDs is not quite accurate. As a rule of thumb, I generally like to keep ¼ of the supply voltage for the series resistor.

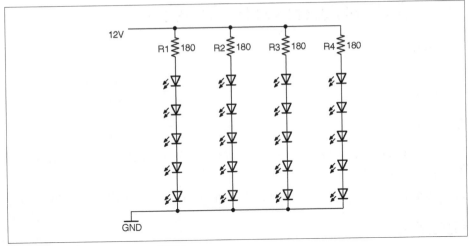

Figure 14-6. Powering Lots of LEDs

Discussion

The extreme version of this design is to power a long series of LEDs directly from mains voltage. It works, but frankly this is just terrifying—don't do it.

It's tempting to consider just putting LEDs in series to share a series resistor as shown in Figure 14-7.

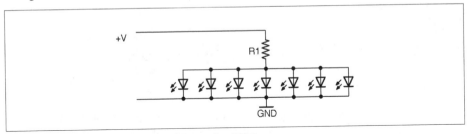

Figure 14-7. An Incorrect Way of Powering LEDs

The problem with doing this is that if there are slight variations in the forward voltage of each LED (which there will be) the LED with the smallest Vf will start conducting first and immediately draw ALL the current from the resistor. This may well burn out the LEDs one at a time.

See Also

For basic information on LEDs and series resistors, see Figure 14-1.

There is a nice online calculator that will do the series resistor calculations for you here: *http://led.linear1.org/led.wiz*.

14.4 Switch Lots of LEDs at the Same Time

Problem

You have a big bank of LEDs (see Recipe 14.3) and you want to switch them all on and off from an Arduino or Raspberry Pi.

Solution

Switching a big bank of LEDs is really no different from switching any sizeable load. You can use Recipe 13.3 to do this.

For programs to turn the LEDs on and off, you can use Recipe 10.8 and Recipe 10.9.

Discussion

You can also use a PWM signal to control the brightness of a big bank of LEDs using the programs found in Recipe 10.13 and Recipe 10.14.

See Also

See Chapter 11 for a whole load of recipes concerned with switching loads like this.

14.5 Multiplex Signals to 7-Segment Displays

Problem

You want to drive a multidigit 7-segment display from an Arduino or Raspberry Pi.

Solution

Use multiplexing—a technique that reduces the number of GPIO pins needed to light lots of LEDs. Figure 14-8 shows how you can control a 4-digit LED display from an Arduino Uno.

Each digit of the display has seven segments in a figure-8 arrangement. Each of these segments is labeled A to G and is connected, inside the LED package, to all the segments of the same name. In this display, all the cathodes of the segments for a particular digit are connected together as a common cathode. The LED display has a pin for each of these four common cathodes so that the software can enable each digit in turn, set its segment pattern, and then move on to the next digit.

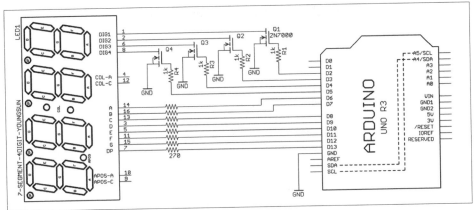

Figure 14-8. Controlling a 4-digit LED Display from an Arduino

The sketch *ch_14_7_seg_mux* displays the numbers 1 to 4 on the display (see Figure 14-9).

Figure 14-9. Multiplexing a 7-Segment Display on a Breadboard

The sketch can be found with the downloads for the book (see Recipe 10.2):

```
const int digitPins[] = {2, 3, 4, 5};
const int segPins[] = {6, 7, 8, 9, 10, 11, 12, 13};

//                      abcdefgD
const char num[] = { 0b11111100,  // 0 abcdef
                     0b00001100,  // 1     ef
                     0b11011010,  // 2 ab de g
                     0b10011110,  // 3 a  defg
                     0b00101110,  // 4   c efg
                     0b10110110,  // 5 a cd fg
                     0b11110110,  // 6 abcd fg
                     0b00011100,  // 7    def
                     0b11111110,  // 8 abcdefg
                     0b10111110}; // 9 a cdefg

int digits[] = {1, 2, 3, 4};

void setup()
{
  for (int i = 0; i < 4; i++)
  {
    pinMode(digitPins[i], OUTPUT);
  }
  for (int i = 0; i < 8; i++)
  {
    pinMode(segPins[i], OUTPUT);
  }
}

void loop()
{
  refreshDisplay();
}

void refreshDisplay()
{
  for (int d = 0; d < 4; d++)
  {
    for (int seg=0; seg < 8; seg++)
    {
      digitalWrite(segPins[seg], LOW);
    }
    digitalWrite(digitPins[d], HIGH);
    for (int seg=0; seg < 8; seg++)
    {
      digitalWrite(segPins[seg], bitRead(num[digits[d]], 7-seg));
    }
    delay(1);
    digitalWrite(digitPins[d], LOW);
  }
}
```

The digit and segment pins are contained in arrays. The `setup` function sets them all as outputs.

Two other arrays are defined: `num` contains the bit patterns for the numbers 0 to 9. A 1 in the bit position means the segment should be lit for that value. The array `digits` holds the numeric value to be displayed on each of the four positions.

Every time the `refreshDisplay` function is called, all four digits of the display will be displayed and then the display blanked. This means that `refreshDisplay` must be called as frequently as possible or the display may become flickery or dim. The function has an outer loop that loops for each digit number d between 0 and 3. Inside this loop, the segments are first turned off and then the control pin for the current digit is enabled.

An inner loop then loops for each segment (`seg`) and determines whether the current segment should be lit or not, based on the expression:

```
bitRead(num[digits[d]], 7-seg)
```

This first finds the current digit value and then looks up the bit pattern for that value.

Discussion

The critical thing about the preceding example code is that anything that you do inside the `loop` function along with `refreshDisplay` must happen quickly, or the display will start to flicker. The human eye will not see the flickering until `refreshDis play` is called less frequently than 30 times a second, giving you a theoretical 30 milliseconds (approximately) to do something else in `loop`. This is more than enough time to check the state of digital inputs and perform other simple actions, but remember—always keep `loop` as fast as possible.

Figure 14-9 shows that there is quite a lot of wiring involved in multiplexing the display. The same hardware can be attached to the GPIO pins of a Raspberry Pi (at least a 40-pin GPIO Raspberry Pi) but the nature of the operating system means that you might find the refresh a little uneven. For a Raspberry Pi, it is probably better to use a display module that has its own hardware such as the I2C LED display in Recipe 14.9.

See Also

To use an I2C 7-segment LED module, see Recipe 14.9.

14.6 Control Many LEDs

Problem

You want to control many LEDs with just a few GPIO pins.

Solution

Use a technique called Charlieplexing.

The name Charlieplexing comes from the inventor Charlie Allen of the company Maxim, and the technique takes advantage of the feature of GPIO pins that allows them to be changed from outputs to inputs while a program is running. When a pin is changed to be an input, not enough current will flow through it to light an LED or influence other pins connected to the LED that are set as outputs.

Figure 14-10 shows the schematic for controlling six LEDs with three pins.

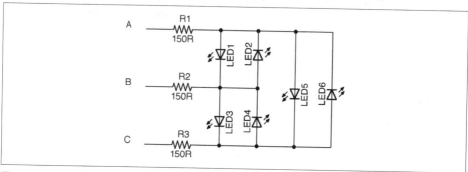

Figure 14-10. Schematic for Charlieplexing Six LEDs

Looking at Figure 14-10, if you want to turn on LED1, you need to make pin A high and B low. But to make sure that none of the other LEDs light, you need to set C to be an input, so that no current can flow into or out of it. More accurately, to make sure that not enough current to light an LED flows into or out of it.

For a small number of LEDs, you can test out Charlieplexing on a breadboard as shown in Figure 14-11.

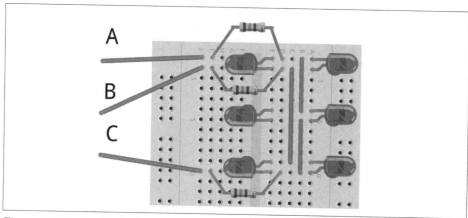

Figure 14-11. Charlieplexing on a Breadboard

Arduino software

Connect the three control points on the breadboard to pins D5, D6, and D7 of your Arduino (D6 to the middle control pin). You can then use the following sketch (*ch_14_charlieplexing*) to try out Charlieplexing. You will find the sketch with the downloads for the book (see Recipe 10.2):

```
const int pins[] = {5, 6, 7};

const int pinLEDstates[6][3] = {
  {1, 0, -1}, // LED 1
  {0, 1, -1}, // LED 2
  {-1, 1, 0}, // LED 3
  {-1, 0, 1}, // LED 4
  {1, -1, 0}, // LED 5
  {0, -1, 1}  // LED 6
};

int ledState[6];

void setup()
{
  Serial.begin(9600);
  Serial.println("LED Number (0 to 5)");
}

void loop()
{
  if (Serial.available())
  {
    int led = Serial.parseInt();
    ledState[led] = ! ledState[led];
  }
  refresh();
}

void refresh()
{

  for (int led = 0; led < 6; led ++)
  {
    clearPins();
    if (ledState[led])
    {
      setPins(led);
    }
    else
    {
      clearPins();
    }
    delay(1);
  }
```

```
    }

    void setPins(int led)
    {
      for (int pin = 0; pin < 3; pin ++)
      {
        if (pinLEDstates[led][pin] == -1)
        {
          pinMode(pins[pin], INPUT);
        }
        else
        {
          pinMode(pins[pin], OUTPUT);
          digitalWrite(pins[pin], pinLEDstates[led][pin]);
        }
      }
    }

    void clearPins()
    {
      for (int pin = 0; pin < 3; pin ++)
      {
          pinMode(pins[pin], INPUT);
      }
    }
```

The key to this code is the pinLEDstates data structure. This specifies the states that the controlling pins should be set to for a particular LED. So, LED3 has the pattern –1, 1, 0. This means that for LED3 to be lit, the first control pin should be set to an input (–1), the second control pin to a HIGH digital output, and the third to a LOW digital output. If you check Figure 14-10 this will confirm those settings.

The loop function first prompts for a particular LED and then toggles that LED turning it on if it was off and off if it was on. The array ledStates is used to keep track of which LEDs should be on or off.

The loop function then calls refresh, which first clears all the pins setting them to inputs using clearPins and then for each of the LEDs uses setPins to either set the pins appropriately to turn the LED on or not depending on that LED's entry in ledState.

This works fine for a fairly small number of LEDs, assuming that the Arduino is not doing much else, so that refresh can be called frequently. You may have noticed that the LEDs flicker when serial communication is in progress through the Serial Monitor.

Raspberry Pi software

Using a Raspberry Pi instead of an Arduino just requires you to swap the male-to-male header leads to be male-to-female connectors and choose three GPIO pins. In the case of the following example program (*ch_14_charlieplexing.py*) these should be 18, 23, and 24. You should also increase the value of the resistors to 270Ω.

You can find the program with the downloads for the book (see Recipe 10.4):

```python
import RPi.GPIO as GPIO
import thread, time


GPIO.setmode(GPIO.BCM)
pins = [18, 23, 24]


pin_led_states = [
  [1, 0, -1], # LED1
  [0, 1, -1], # LED2
  [-1, 1, 0], # LED3
  [-1, 0, 1], # LED4
  [1, -1, 0], # LED5
  [0, -1, 1]  # LED6
]

led_states = [0, 0, 0, 0, 0, 0]

def set_pins(led):
  for pin in range(0, 3):
      if pin_led_states[led][pin] == -1:
          GPIO.setup(pins[pin], GPIO.IN)
      else:
          GPIO.setup(pins[pin], GPIO.OUT)
          GPIO.output(pins[pin], pin_led_states[led][pin])

def clear_pins():
  for pin in range(0, 3):
      GPIO.setup(pins[pin], GPIO.IN)

def refresh():
  while True:
    for led in range(0, 6):
      clear_pins()
      if led_states[led]:
        set_pins(led)
      else:
        clear_pins()
      time.sleep(0.001)

thread.start_new_thread(refresh, ())
```

```
while True:
    x = int(raw_input("Pin (0 to 5) :"))
    led_states[x] = not led_states[x]
```

The Raspberry Pi version follows the same pattern as the Arduino version except that the call to refresh is in a separate thread of execution, so the display will automatically update, even if the Python program is waiting for input.

Discussion

The number of LEDs that can be controlled for the number of GPIO pins (n) used is given by the formula:

$$leds = n^2 - n$$

Using four pins, you can have 16–4 = 12 LEDs, whereas 10 pins would give you a massive 90 LEDs.

See Also

For a great description of Charlieplexing as well as layouts for large numbers of LEDs, see *https://en.wikipedia.org/wiki/Charlieplexing*.

14.7 Change the Colors of RGB LEDs

Problem

You want to set the color of an RGB LED connected to the GPIO pins of a Raspberry Pi or Arduino.

Solution

Wire up a common-cathode RGB LED as shown in Figure 14-12.

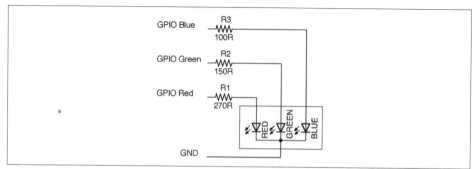

Figure 14-12. Wiring up an RGB LED

The red, green, and blue LEDs, in this example, all have different values of series resistor to try and provide equal brightness. The three LED channels can either be simply turned on and off to mix seven different colors from the three control pins, or you can use PWM on the pins to mix more subtle colors.

Identifying Pins of an RGB LED

Figure 14-13 shows a typical RGB LED identifying the pins.

By convention (but check your datasheet) the longest lead is the common lead and the other three leads are each for one color channel.

If you are not sure about a device, attach a 1k resistor to the positive end of a voltage source such as a lab power supply and 9V battery and then test all combinations of the connections. First, identify the common connection (anode or cathode) and then work out which color is controlled by which lead.

Figure 14-13. RGB LED Pins

Arduino software

Connect an RGB LED to an Arduino as shown in Figure 14-12 using the three Arduino pins 9, 10, and 11 for the blue, green, and red channels, respectively.

The following sketch can be found with the downloads for the book (Recipe 10.2) and is called *ch_14_rgb_led*:

```
const int redPin = 11;
const int greenPin = 10;
const int bluePin = 9;

void setup() {
  pinMode(redPin, OUTPUT);
  pinMode(greenPin, OUTPUT);
  pinMode(bluePin, OUTPUT);
  Serial.begin(9600);
  Serial.println("Enter R G B (E.g. 255 100 200)");
}

void loop() {
  if (Serial.available()) {
    int red = Serial.parseInt();
    int green = Serial.parseInt();
    int blue = Serial.parseInt();
    analogWrite(redPin, red);
    analogWrite(greenPin, green);
    analogWrite(bluePin, blue);
  }
}
```

By sending three numbers separated by spaces from the Arduino Serial Monitor, you can mix pretty much any color from the LED.

Raspberry Pi software

When using a Raspberry Pi, you can follow the same approach of creating three PWM channels using the RPi.GPIO library (see Recipe 10.14) or, as in the following example, you can install the Python Squid library, which simplifies the process.

To install the Squid library, run the following commands:

```
$ git clone https://github.com/simonmonk/squid.git
$ cd squid
$ sudo python setup.py install
```

The equivalent program for controlling the RGB LED from a Raspberry Pi can be a little fancier and use the Tkinter Python library to make a user interface that allows you to use sliders to adjust the red, green, and blue channels (Figure 14-14):

```
from squid import *
from Tkinter import *

rgb = Squid(18, 23, 24)

class App:
```

```python
    def __init__(self, master):
        frame = Frame(master)
        frame.pack()
        Label(frame, text='Red').grid(row=0, column=0)
        Label(frame, text='Green').grid(row=1, column=0)
        Label(frame, text='Blue').grid(row=2, column=0)
        scaleRed = Scale(frame, from_=0, to=100,
            orient=HORIZONTAL, command=self.updateRed)
        scaleRed.grid(row=0, column=1)
        scaleGreen = Scale(frame, from_=0, to=100,
            orient=HORIZONTAL, command=self.updateGreen)
        scaleGreen.grid(row=1, column=1)
        scaleBlue = Scale(frame, from_=0, to=100,
            orient=HORIZONTAL, command=self.updateBlue)
        scaleBlue.grid(row=2, column=1)
    def updateRed(self, duty):
        rgb.set_red(float(duty))
    def updateGreen(self, duty):
        rgb.set_green(float(duty))

    def updateBlue(self, duty):
        rgb.set_blue(float(duty))

root = Tk()
root.wm_title('RGB LED Control')
app = App(root)
root.geometry("200x150+0+0")
root.mainloop()
```

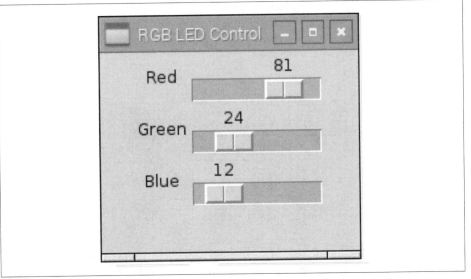

Figure 14-14. Adjusting the RGB LED Color Using a User Interface

Discussion

The LED in Figure 14-12 is a common cathode device; that is, all the negative terminals (cathodes) of the LEDs are connected together. RGB LEDs are also available in common anode form, which can sometimes be more convenient if they are being controlled by transistors as it allows low-side switching.

See Also

For recipes on PWM for Arduino and Raspberry Pi, see Recipe 10.13 and Recipe 10.14, respectively.

The easiest way to control large numbers of RGB LEDs is to use addressable LED strips as described in Recipe 14.8.

14.8 Connect to Addressable LED Strips

Problem

You want to control a strip of addressable LEDs, often referred to as neopixels, from an Arduino or Raspberry Pi.

Solution

Do the power calculations carefully for your pixels and then use a single GPIO pin to send the data to the pixel array. Figure 14-15 shows a typical arrangement for a large number of pixels. If you only have a few pixels (say 5–10), you can power them directly from your Arduino or Raspberry Pi.

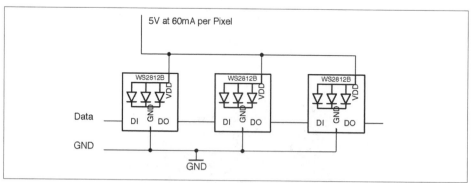

Figure 14-15. Schematic for Using WS2812 Addressable LED Strips

If you don't mind taking the risk of damaging your Pi, Arduino, or whatever is supplying USB power to them, you can power more pixels as long as you are extremely careful in your code to keep the brightness of the pixels well below the maximum brightness. However, this is not recommended.

Arduino software

To test out an addressable LED strip with an Arduino, connect pin D9 to the data terminal of an addressable LED strip.

The Arduino sketch *ch_14_neopixel* is included in the downloads for the book (see Recipe 10.2). The sketch uses the Adafruit NeoPixel library. This can be installed from within the Arduino IDE using the Library Manager by selecting the menu option Sketch→Include Library→Manage Libraries.

Once the Library Manager is open scroll down and select Adafruit NeoPixel and then click Install.

```
#include <Adafruit_NeoPixel.h>

const int pixelPin = 9;
const int numPixels = 10;

Adafruit_NeoPixel pixels = Adafruit_NeoPixel(numPixels, pixelPin,
                                    NEO_GRB + NEO_KHZ800);

void setup() {
  pixels.begin();
}

void loop() {
  for (int i = 0; i < numPixels; i++) {
    int red = random(64);
    int green = random(64);
    int blue = random(64);
    pixels.setPixelColor(i, pixels.Color(red, green, blue));
    pixels.show();
  }
  delay(100);
}
```

Change the values of `pixelPin` and `numPixels` to match the GPIO pin you are using and the number of pixels in your strip.

Each pixel will be allocated red, green, and blue intensities at random. Note that the maximum of 64 is used rather than the full range of 0 to 255 because an intensity of 255 is actually very bright.

Raspberry Pi software

To test out an addressable LED strip with Raspberry Pi, connect GPIO10 to the data terminal of an addressable LED strip.

To use the display, you first need to install some libraries by running the following commands:

```
$ git clone https://github.com/doceme/py-spidev.git
$ cd py-spidev/
$ make
$ sudo make install
$ cd ..
$ git clone https://github.com/joosteto/ws2812-spi.git
$ cd ws2812-spi
$ sudo python setup.py install
```

You still also need to enable SPI on your Raspberry Pi (see Recipe 10.16).

You can find a program to test out the LED display in the file *ch_14_neopixels.py*. See Recipe 10.4 for information on downloading the Python example programs.

```
import spidev
import ws2812
from random import randint
import time

spi = spidev.SpiDev()
spi.open(0,0)

N = 10

                            # g r b
pixels = []
for x in range(0, 10):
    pixels.append([0, 0, 0])

while True:
  for i in range(0, N):
    pixels[i] = [randint(0, 64), randint(0, 64), randint(0, 64)]
    ws2812.write2812(spi, pixels)
  time.sleep(0.1)
```

Each pixel is represented in an array of three values. These are in order green, red, and blue rather than the more usual red, green, blue. That's just how the library is. An array (pixels) of N elements where N is the number of pixels is created by using a for loop to append N arrays of green, red, and blue to the pixels array.

The main loop acts just like its Arduino counterpart. It allocates color intensities at random to each pixel position before using the ws2812 library to write out the pixels to the GPIO pin.

Discussion

If you are using a Raspberry Pi, then you may need to level shift the data output (Recipe 10.17), although I have never found this to be necessary in practice. Looking at the datasheet for the WS2812, the minimum input voltage to count as a logical HIGH is 0.7 times the supply voltage (5V). This gives a theoretical value of 3.5V.

You can make a convenient display to attach to an Arduino or Raspberry Pi by sacrificing jumper leads by snipping off one connector and soldering the wire to the LED strip. Figure 14-16 shows such a display connected to a Raspberry Pi. Note the use of the heat-shrink sleave to help stop the solder joints from flexing.

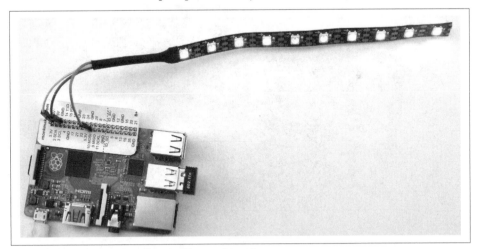

Figure 14-16. A Handy Addressable Pixel Display

See Also

For more information on the ws2812-spi library, see *https://github.com/joosteto/ws2812-spi*, including instructions on using the NumPy library to improve performance for long LED strings.

You can find the datasheet for the WS2812 here: *https://cdn-shop.adafruit.com/data-sheets/WS2812.pdf*.

To control a single RGB LED directly, see Recipe 14.7.

Addressable LED pixels are not confined to strips of LEDs; you can also buy them organized into rings (Adafruit product 1586) and arrays (Adafruit product 1487).

An alternative library that is based on the Raspberry Pi's DMA hardware rather than its SPI can be found here: *https://github.com/richardghirst*.

14.9 Use an I2C 7-Segment LED Display

Problem

You want to use a 7-segment display but without the spaghetti wiring of Recipe 14.5.

Solution

Use a ready-made I2C display module like the one shown in Figure 14-17.

Figure 14-17. An Adafruit 7-Segment Display

This module and similar modules that can be found on eBay, with four or even eight digits, are a convenient way to add a 7-segment LED display to an Arduino or Raspberry Pi.

Figure 14-18 shows how to wire one up using two GPIO pins.

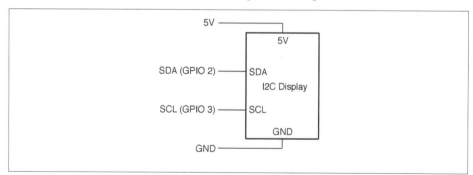

Figure 14-18. Connecting an I2C Display to an Arduino or Raspberry Pi

The I2C serial interface needs two data pins as well as 5V of power. On both the Arduino and Raspberry Pi, specific pins must be used for the I2C interface. On the

Arduino Uno, these are the SCL (serial clock) and SDA (serial data) pins (see Recipe 10.7) and on a Raspberry Pi, these are GPIO2 and GPIO3.

This display and many types of displays like it use the HT16K33 IC, which can actually control up to 16 segments in 8 digits. Although the module needs to be powered with 5V, its I2C interface works fine from 3.3V Raspberry Pi GPIO pins, so no level conversion is necessary.

Arduino software

The sketch uses the Adafruit LED Backpack library and Adafruit GFX library. These can be installed from within the Arduino IDE using the Library Manager by selecting the menu option Sketch→Include Library→Manage Libraries.

Once the Library Manager has opened scroll down and select the two libraries and and then click Install.

Adafruit provides a whole load of examples for using the library that you can access directly from the Arduino IDE. A good example to try out can be found under the menu item File→Examples→Adafruit Backpack Library→sevenseg.

Raspberry Pi software

The connections between the Raspberry Pi and the module are as follows:

- VCC (+) on the display to 5V on the Raspberry Pi GPIO connector
- GND (–) on the display to GND on the Raspberry Pi GPIO connector
- SDA (D) on the display to GPIO 2 (SDA) on the Raspberry Pi GPIO connector
- SCL (C) on the display to GPIO 3 (SCL) on the Raspberry Pi GPIO connector

VCC is the abbreviation for *volts collector collector*, and is frequently used to identify the positive power supply pin of an IC or module.

Before you can use the display, you will need to enable the Raspberry Pi's I2C interface by following Recipe 10.15.

Adafruit also provides Python code for this module. To install it, enter the following commands into the terminal:

```
$ sudo apt-get update
$ sudo apt-get install build-essential python-dev
$ sudo apt-get install python-imaging
$ git clone https://github.com/adafruit/Adafruit_Python_LED_Backpack.git
$ cd Adafruit_Python_LED_Backpack
$ sudo python setup.py install
```

A good example to try out is the clock example *sevensegment_test.py*.

Discussion

Building a project using a module like this is fine when you are developing a proto-type, but in a finished product, you would most likely use direct multiplexing as described in Recipe 14.5 (to reduce costs) or use the HT16K33 IC on a circuit board of your own design, if you really need a hardware driver.

I2C Addresses

In theory, you can connect up to 255 devices using the same two SDA and SCL pins of an Arduino or Raspberry Pi. Each device attached to the bus must have its own unique address.

This means that if you have, say, two identical displays connected to the same bus, you need to be able to change the address of one of the modules. On most I2C devi-ces, you will find solder switches like the ones shown in Figure 14-19 that allow you to set a certain address by bridging the gaps with solder. You should refer to the docu-mentation for the module to do this.

Figure 14-19. Using Solder Switches to Select the Address of an I2C Module

See Also

You can find lots more information on this display at its project page on Adafruit (*https://www.adafruit.com/products/881*).

To control a similar display but using multiplexing, see Recipe 14.5.

You can download the HT16K33 datasheet from *http://bit.ly/2mbaWyP*.

14.10 Display Graphics or Text on OLED Displays

Problem

You need to display text and graphics on a small display.

Solution

Use an I2C OLED display like the one shown attached to an Arduino in Figure 14-20.

Figure 14-20. An I2C OLED Display Attached to an Arduino

The display has the same four connections as the display used in Recipe 14.9 so you can make the same connections as shown in Figure 14-18.

Arduino software

Wire up the display as described in Recipe 14.9.

You will also need to add the Adafruit GFX and SSD1306 libraries to your Arduino IDE by using the Library Manager (from the menu Sketch→Include Library→Library

Manager and then scroll down the list and install both the Adafruit GFX Library and the Adafruit SSD1306 library).

The example sketch *ch_14_oled* will display the message shown in Figure 14-20. To install the Arduino programs for the book, see Recipe 10.2.

```
#include <Wire.h>
#include <Adafruit_GFX.h>
#include <Adafruit_SSD1306.h>

Adafruit_SSD1306 display(4);

void setup()
{
  display.begin(SSD1306_SWITCHCAPVCC, 0x3c);
  display.clearDisplay();
  display.drawRect(0, 0, display.width()-1, display.height()-1, WHITE);
  display.setTextSize(1);
  display.setTextColor(WHITE);
  display.setCursor(5,10);
  display.print("Electronics Cookbook");
  display.display();
}

void loop()
{
}
```

Raspberry Pi software

Wire up the display as described in Recipe 14.9.

Before you can use the display, you will need to enable the Raspberry Pi's I2C interface by following Recipe 10.15.

You will also need to download and install the SSD1306 Python library and prerequisites using the following commands:

```
$ sudo pip install pillow
$ git clone https://github.com/rm-hull/ssd1306.git
$ cd ssd1306
$ sudo python setup.py install
```

Now change the directory to the book's example programs and run the program *ch_14_oled.py* (see Recipe 10.4). The text "Electronics Cookbook" with a rectangular border should appear on the display. If it doesn't, then there is a good chance you will need to change the I2C address of the device in the file *demo_opts.py* that is with the other downloads. Change line 13 from 0x3c to the I2C address of your device. The file *ch_14_oled.py* is listed here:

```
from demo_opts import device
from oled.render import canvas
```

```
from PIL import ImageFont
from demo_opts import args

font = ImageFont.load_default()

with canvas(device) as draw:
    draw.rectangle((0, 0, device.width-1, device.height-1), outline=255, fill=0)
    font = ImageFont.load_default()
    draw.text((5, 20), 'Electronics Cookbook',  font=font, fill=255)
```

Discussion

Once you get into the realm of needing a display that can show text and graphics, you can of course use an HDMI monitor with your Raspberry Pi.

See Also

For a simple numeric display see Recipe 14.9 and for a low-cost two-line alphanumeric display see Recipe 14.11.

The SSD1306 library page on GitHub has more examples and documentation for using these displays: *https://github.com/rm-hull/ssd1306*.

14.11 Display Text on Alphanumeric LCD Displays

Problem

You need a low-cost alphanumeric display.

Solution

Use an LCD module based on the HD44780 IC. Figure 14-21 shows one of these displays wired to an Arduino and Figure 14-22 shows how the display should be connected on a breadboard. Figure 14-23 shows the connections for a Raspberry Pi as a schematic.

The HD44780 can be configured to use either an 4- or 8-bit parallel data bus. If the 4-bit bus is used only bits 4 to 7 of the bus are used. The pin Vo is used to control the contrast of the screen. You will need to adjust R1 to be able to see anything on the screen.

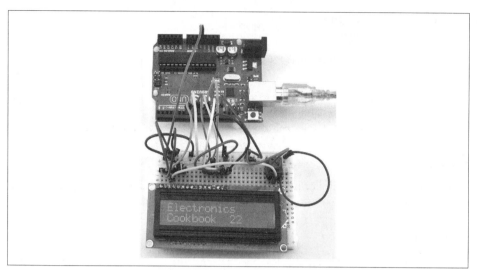

Figure 14-21. An HD44780 16x2 LCD Display Connected to an Arduino

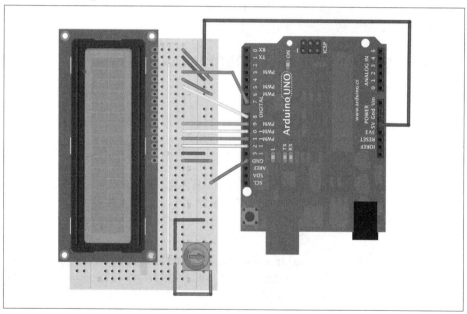

Figure 14-22. Wiring an HD44780 Display to an Arduino Uno (Breadboard)

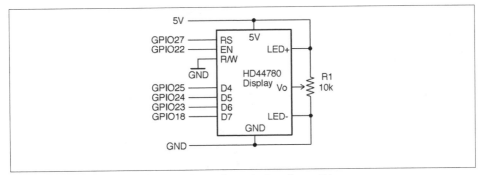

Figure 14-23. Wiring an HD44780 Display to a Raspberry Pi

Arduino software

The Arduino IDE includes a library called LiquidCrystal that takes care of all the communications with the HD44780 IC. The following example can be found with the Arduino downloads for the book (Recipe 12.2) in the sketch *ch_14_lcd*:

```
#include <LiquidCrystal.h>

//              RS EN D4 D5  D6  D7
LiquidCrystal lcd(7, 8, 9, 10, 11, 12);

void setup() {
  lcd.begin(16, 2);
  lcd.print("Electronics");
  lcd.setCursor(0, 1);
  lcd.print("Cookbook");
}

void loop() {
  lcd.setCursor(10, 1);
  lcd.print(millis() / 1000);
}
```

Raspberry Pi software

To use an HD44780 display on a Raspberry Pi, you will first need to install the Adafruit CharLCD Python library by entering the following commands:

```
$ git clone https://github.com/adafruit/Adafruit_Python_CharLCD.git
$ cd Adafruit_Python_CharLCD
$ sudo python setup.py install
```

You can now try out the example program *ch_14_lcd.py*:

```
import time
import Adafruit_CharLCD as LCD

# Raspberry Pi pin configuration:
```

```
lcd_rs        = 27  # Note this needs to be changed to 21 for Model B rev1 Pis.
lcd_en        = 22
lcd_d4        = 25
lcd_d5        = 24
lcd_d6        = 23
lcd_d7        = 18
lcd_backlight = 4

lcd_columns = 16
lcd_rows    = 2

lcd = LCD.Adafruit_CharLCD(lcd_rs, lcd_en, lcd_d4, lcd_d5, lcd_d6, lcd_d7,
                          lcd_columns, lcd_rows, lcd_backlight)

lcd.message('Electyronics\nCookbook')
t0 = time.time()

while True:
    lcd.set_cursor(10, 1)
    lcd.message(str(int(time.time()-t0)))
    time.sleep(0.1)
```

Discussion

These displays are available in different sizes. So, in addition to the 16x2 (2 rows of 16 characters) used above, other common sizes are 8x1, 20x2, and 20x4.

See Also

Adafruit has a wide selection of these types of displays, including ones that have an RGB color backlight (*https://www.adafruit.com/products/399*).

## CHAPTER 15

Digital ICs

15.0 Introduction

In many projects the only digital IC you will need is a microcontroller. With sufficient GPIO pins you can do most digital tasks. In this chapter, you will find recipes for digital ICs that still manage to find a role in modern electronics design.

Sometimes, a single digital logic IC is all you need. A single digital logic IC is often cheaper and removes the need for programming that would otherwise arise if you used a microcontroller.

15.1 Protecting ICs from Electrical Noise

Problem

You want to use an IC, but you want to avoid errors caused by electrical noise.

Solution

Use a 100nF capacitor as close as possible to the power pins of the digital IC. The shorter the leads to the capacitor the better.

This process of connecting a capacitor with short circuit board tracks to the power pins of an IC is called "decoupling" because each IC has its own little reservoir of charge and will therefore not be "coupled" to a neighboring capacitor by affecting its power supply. Capacitors used in this way are also called "bypass" capacitors.

Figure 15-1 shows a circuit board containing a digital IC with a 100nF multilayer ceramic (MLC) capacitor for both a through-hole and SMD.

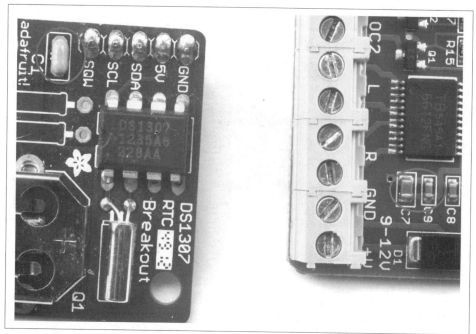

Figure 15-1. Decoupling Capacitors

The circuit board on the right of Figure 15-1 actually has two capacitors in parallel as the decoupling capacitor. The smaller one is 100nF and the larger one 10μF. You may think that the effect of the 100nF would be lost, but its lower ESR (see Recipe 3.2) allows it to respond faster to pulses of power consumption from the chip. So, the 10μF capacitor is in turn providing a larger local energy store for the smaller capacitor and IC to draw on, but one that has a higher ESR and therefore cannot respond as quickly. Such an arrangement of ever-decreasing capacitor values as you approach the power rails of the IC is particularly common where larger currents are being switched such as in motor controllers and audio power amplifiers.

Discussion

Digital ICs contain lots of transistors that switch very quickly. The decoupling capacitor provides a little reservoir of charge so that this switching does not result in excessive electrical noise on the power lines that could go on to affect other parts of the circuit.

It is considered to be good practice to include a decoupling capacitor next to every digital IC. In fact, this is also true of analog ICs.

See Also

For more information on capacitors, see Chapter 3.

15.2 Know Your Logic Families

Problem

You want to know what type of logic family you should use and its characteristics.

Solution

Use high-speed CMOS (complimentary metal-oxide semiconductor), the 74HC family of chips, unless you are repairing vintage electronics.

Discussion

Logic-gate ICs, like the mafia, are arranged into families. TTL (transistor transistor logic) used to be a popular choice, but is now totally obsolete unless you are repairing an equally obsolete computer from the last century. TTL chips starting with the number 74 (e.g., 7400) used to compete with CMOS chips with names that started with 40 (e.g., 4011). Each had their own overlapping ranges of logic gates, flip-flops, shift registers, and counters.

TTL was faster than CMOS, but CMOS used less current and was far less fussy about its supply-voltage range. Now these two families have merged into a single "best-of-both-worlds" family called high-speed CMOS. If you started playing with electronics in the 1970s, you can still find your favorite logic ICs from both families but under one name starting with 74HC. For instance, the 7400 of old is now the 74HC00 and the 4011 of old is now called the 74HC4001.

A high-speed CMOS chip has a supply-voltage range of 2V to 6V and only consumes about 1μA until it starts switching. Their outputs can also generally source or sink 4mA at their outputs (per gate).

The original 40xx CMOS is still available and can be useful where you need a greater power-supply range than high-speed CMOS can provide.

See Also

For the datasheet of a typical high-speed CMOS device, see *http://www.ti.com/lit/ds/symlink/sn74hc00.pdf*.

15.3 Control More Outputs Than You Have GPIO Pins

Problem

You have run out of GPIO pins on your Arduino or Raspberry Pi and want to control some LEDs.

Solution

Use a serial-to-parallel shift register such as the 74HC4094 and write some code to load its registers with data using a serial interface that only requires three pins. LEDs can then be attached to the outputs. Figure 15-2 shows a schematic for this.

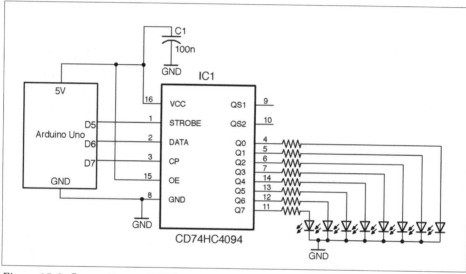

Figure 15-2. Connecting an 74HC4094 to an Arduino Uno

The total current allowed by the 74HC4094 is 50mA so assuming 6mA per LED, a suitable series resistor would be 680Ω if all the LEDs are lit at the same time, to allow 8 x 6mA = 48mA.

Arduino software

You can find the Arduino code to send data to the shift register with the downloads for the book (Recipe 10.2) in the sketch *ch_15_sift_reg*:

```
const int strobePin = 5;
const int dataPin = 6;
const int clockPin = 7;

void setup() {
```

```
    pinMode(strobePin, OUTPUT);
    pinMode(dataPin, OUTPUT);
    pinMode(clockPin, OUTPUT);
    Serial.begin(9600);
    Serial.println("Enter Byte");
}

void loop() {
    if (Serial.available()) {
        char bits = Serial.parseInt();
        shiftOut(dataPin, clockPin, MSBFIRST, bits);
        digitalWrite(strobePin, HIGH);
        delayMicroseconds(10);
        digitalWrite(strobePin, LOW);
        Serial.println(bits, 2);
    }
}
```

This sketch uses the Arduino `shiftOut` function to send the serial data and uses these parameters: the pin to send the data on, the clock pin, a flag to determine the order the data is sent (in this case, most significant bit [MSB] first), and the actual data to send.

If you open the Arduino Serial Monitor, you will be prompted for a value to load into the shift register. This value will be shifted in and then a confirmation of the data in binary will be displayed as shown in Figure 15-3. The value entered should be a decimal value between 0 and 255. This value will be echoed in binary in the Serial Monitor for confirmation.

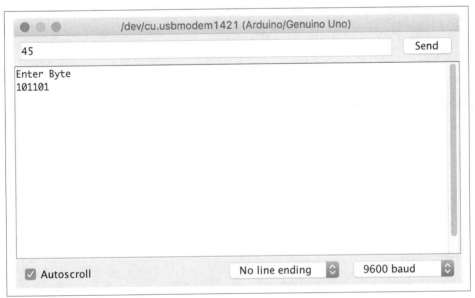

Figure 15-3. Sending Data to the Shift Register

Raspberry Pi software

The following test program assumes you have connected the STROBE pin of the 74HC4094 to GPIO18, the DATA pin to GPIO23, and the CLOCK pin to GPIO24.

The Python code (*ch_15_shift_reg.py*) follows a similar pattern as that of the Arduino, but in this case we have to implement the shift_out function. This sets the data_pin to the eighth bit of the data, and then pulses the clock_pin before shifting the data left by one bit position, and so on for all eight bits in MSB first order:

```python
import RPi.GPIO as GPIO
import time

GPIO.setmode(GPIO.BCM)

strobe_pin = 18
data_pin = 23
clock_pin = 24

GPIO.setup(strobe_pin, GPIO.OUT)
GPIO.setup(data_pin, GPIO.OUT)
GPIO.setup(clock_pin, GPIO.OUT)

def shift_out(bits): # MSB first. 8 bits
    for i in range(0, 8):
        b = bits & 0b10000000
        bits = bits << 1
        GPIO.output(data_pin, (b == 0b10000000))
        time.sleep(0.000001)
        GPIO.output(clock_pin, True)
        time.sleep(0.000001)
        GPIO.output(clock_pin, False)
        time.sleep(0.000001)

try:
    while True:
        bits = input("Enter Byte ")
        print(bin(bits))
        shift_out(bits)
        GPIO.output(strobe_pin, True)
        time.sleep(0.000001)
        GPIO.output(strobe_pin, False)

finally:
    print("Cleaning up")
    GPIO.cleanup()
```

Discussion

Figure 15-4 shows the logical diagram for the 74HC4094.

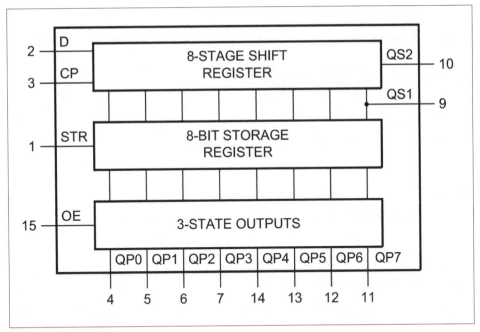

Figure 15-4. The 74HC4094

The 74HC4094 is a serial-to-parallel shift register. The 8-stage shift register's data is set a bit at a time by setting the data (D) pin either HIGH or LOW and then pulsing the clock (CP) pin to transfer that bit. The data pin is then set to the next value and the clock pulsed again to shift the bit already in the shift register along one position and then add the new bit at the beginning. This process continues until all eight bits are loaded into the shift register.

The contents of the shift register are not actually transferred to the 74HC4094's outputs until the strobe pin (STR) is pulsed.

The 74HC4094 also has an output enable (OE) pin that switches all eight outputs either to a high impedance state or to the bit pattern of HIGHs and LOWS in the 3-state-outputs register. If you connect this OE pin to a PWM output you can use it to control the brightness of all the LEDs at the same time.

You can cascade a number of shift registers together by tying their clocks together and attaching the QS2 data output from one shift register to the data input of the next.

See Also

Another way of controlling lots of LEDs from a small number of GPIO pins is to use Charlieplexing (Recipe 14.6).

You can find the datasheet for the 74HC4094 here: *http://bit.ly/2mqB0ET*.

For more information on the Arduino shiftOut function, see *http://bit.ly/2msRHAg*.

15.4 Build a Digital Toggle Switch

Problem

You want to replace a toggle switch with two push switches that turn an LED on and off.

Solution

Use an 74HC00 IC to make a reset-set flip-flop as shown in Figure 15-5.

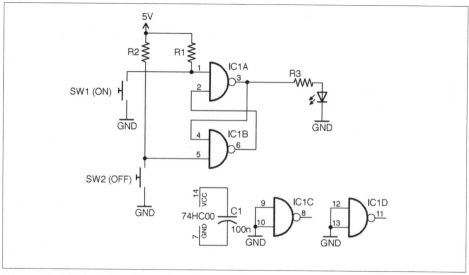

Figure 15-5. Using Flip-Flop and Push Buttons to Switch an LED On and Off

When SW1 is pressed the LED will light and stay lit until SW2 is pressed.

Discussion

R1 and R2 are pull-up resistors that keep the inputs to flip-flop high until the buttons are pressed. One advantage of this way of switching is that if the switches bounce (Recipe 12.1) it will have no effect on the operation of the circuit.

Note that in Figure 15-5 the inputs of the two unused gates in the 74HC00 are tied to ground. This is considered good practice as it prevents the otherwise floating inputs from causing the gates to oscillate in time with electrical noise.

In addition to switching an LED on or off, you could adapt this circuit to switch large loads with a transistor by combining it with Recipe 11.1 or Recipe 11.3.

See Also

You can find the datasheet for the 74HC00 here: *http://bit.ly/2lLK0R5*.

15.5 Reduce a Signal's Frequency

Problem

You have a high frequency that you want to reduce to a lower frequency.

Solution

Use a frequency-divider IC such as the 8-stage 74HC590 in the arrangement shown in Figure 15-6.

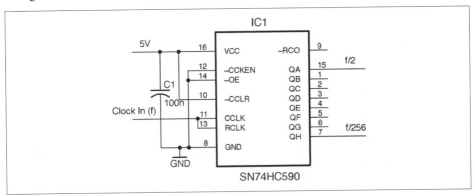

Figure 15-6. A Frequency Divider Using the 74HC590

The output at QA will be half the input frequency f, QB a quarter, and so on until QH has an output frequency of 1/256 of the input frequency.

Discussion

For most designs it is preferable to use a microcontroller when it comes to things like counting. However, counting clock cycles with a microcontroller is significantly slower than the clock frequency of the microcontroller itself. So, a maximum frequency might be a few hundred kHz. To count higher frequencies, QH or one of the other outputs of the 74HC590 could be connected to a digital input on a microcontroller to allow frequencies up to the 24MHz maximum the 74HC590 allows.

See Also

You can find the datasheet for the 74HC590 here: *http://www.nxp.com/documents/data_sheet/74HC590.pdf.*

15.6 Connect to Decimal Counters

Problem

You have run out of GPIO pins on your Arduino or Raspberry Pi and want 10 more outputs of which only one is HIGH at a time.

Solution

Connect the clock and reset the pins of a 74HC4017 decimal counter IC to an Arduino or Raspberry Pi and attach LEDs to the outputs of the 74HC4017 with suitable current-limiting resistors as shown in Figure 15-7.

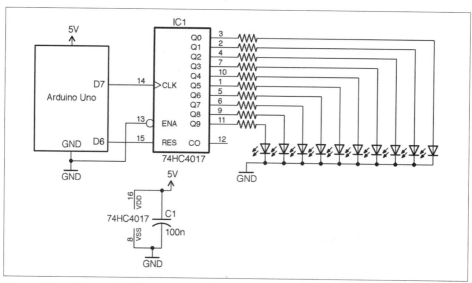

Figure 15-7. Using a 74HC4017 Decimal Counter with an Arduino

The 74HC4017 is a decade counter with decoded outputs. That is, each time it receives a pulse on its clock (CLK) pin it advances to the next output. So, first Q0 is high, then Q1, and so on. The reset pin (RES) sets the counter back to the first output Q0.

Since only one of the LEDs is ever on at a time, the full 20mA output of the 74HC4017 can be used to drive the LEDs.

Arduino software

The Arduino sketch *ch_15_decade_counter* is given here. You can find it with the downloads for the book (Recipe 10.2):

```
const int resetPin = 6;
const int clockPin = 7;

void setup() {
  pinMode(resetPin, OUTPUT);
  pinMode(clockPin, OUTPUT);
  Serial.begin(9600);
  Serial.println("Enter digit 0..9");
}

void loop() {
  if (Serial.available()) {
    int digit = Serial.parseInt();
    setDigit(digit);
  }
}

void setDigit(int digit) {
  digitalWrite(resetPin, HIGH);
  delayMicroseconds(1);
  digitalWrite(resetPin, LOW);
  for (int i = 0; i < digit; i++) {
    digitalWrite(clockPin, HIGH);
    delayMicroseconds(1);
    digitalWrite(clockPin, LOW);
    delayMicroseconds(1);
  }
}
```

The code starts by pulsing the reset pin so that the output Q0 will be high. It then issues the number of pulses to the clock pin requested by the number sent from the Serial Monitor.

The reason the LEDs don't flicker and you don't see the LEDs before the selected LED flash is that this pulsing happens very quickly (in fact, in a few millionths of a second).

Raspberry Pi software

The equivalent Python sketch for the Raspberry Pi assumes you have GPIO18 connected to the reset pin of the 74HC4017 and GPIO23 connected to the clock pin.

The program is called *ch_15_decade_counter.py*. See Recipe 10.4 for information on installing the example programs:

```
import RPi.GPIO as GPIO
import time
```

```
GPIO.setmode(GPIO.BCM)

reset_pin = 18
clock_pin = 23

GPIO.setup(reset_pin, GPIO.OUT)
GPIO.setup(clock_pin, GPIO.OUT)

def set_digit(digit):
    GPIO.output(reset_pin, True)
    time.sleep(0.000001)
    GPIO.output(reset_pin, False)
    time.sleep(0.000001)
    for i in range(0, digit):
        GPIO.output(clock_pin, True)
        time.sleep(0.000001)
        GPIO.output(clock_pin, False)
        time.sleep(0.000001)

try:
    while True:
        digit = input("Enter digit 0..9 ")
        set_digit(digit)

finally:
    print("Cleaning up")
    GPIO.cleanup()
```

The program works in just the same way as its Arduino counterpart.

Discussion

Using a counter like the 74HC4017 can be useful when multiplexing an LED display (Recipe 14.5), as you can use it to select each digit of a 7-segment display or column of an LED matrix, without tying up too many GPIOs.

See Also

You can find the datasheet for the 74HC4017 here: *http://www.nxp.com/documents/ data_sheet/74HC_HCT4017.pdf*.

## CHAPTER 16

Analog

16.0 Introduction

This chapter builds on some of the fundamental principles of resistors, capacitors, and transistors that you have learned about earlier in the book. It also introduces the extremely useful and versatile 555 timer IC.

The analog theme will be continued in Chapters 17, 18, and 19.

16.1 Filter Out High Frequencies (Quick and Dirty)

Problem

You want to filter out the high-frequency component of a signal, for instance, to convert a pulsed digital output into a smooth analog output using a low-pass filter.

Solution

In these circumstances, a simple RC filter (Figure 16-1) will suffice to remove most of the unwanted high-frequency PWM carrier.

Intuitively, all the resistor and capacitor do is make the output slow to respond to changes to the input. All that remains is to decide on suitable values for R and C. We can illustrate this with an example.

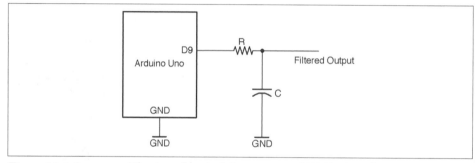

Figure 16-1. Low-pass Filtering of an Arduino-generated PWM Signal

The Arduino Mozzi library (Recipe 18.1) generates PWM (Recipe 10.13) audio output. The length of the pulses on the constant PWM frequency of 32.7kHz determine the amplitude of the lower frequency underlying the audio signal (440 Hz). Figure 16-2 illustrates this.

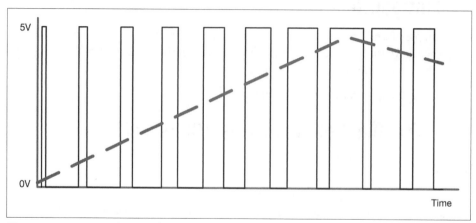

Figure 16-2. PWM of an Audio Signal

The average output voltage is shown by the dashed line, which increases as the pulses lengthen.

Using a value of 270Ω for R and \330nF for C produces the filtering shown in Figure 16-3. The top trace shows the filtered output and the bottom trace the PWM signal.

You can see that the original sine wave from (top) has been extracted from the PWM signal (bottom).

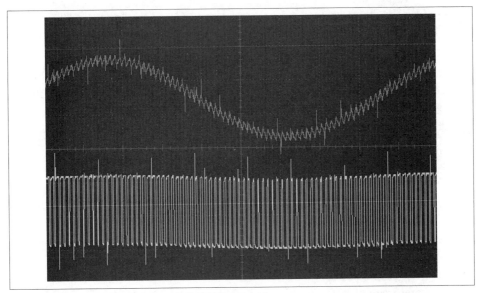

Figure 16-3. Low-pass Filtering of a PWM Signal

Discussion

The kind of RC filter used here is called a first-order filter and only really works well if the signal being filtered is much greater than the frequency you want to keep. Figure 16-4 shows the characteristic plot of gain against frequency.

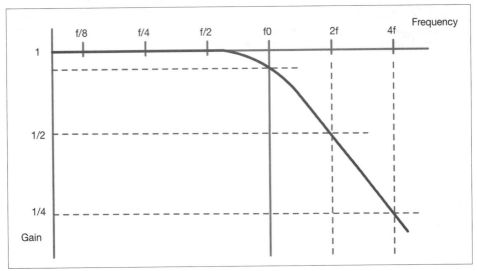

Figure 16-4. A Low-pass Filter's Frequency Response

A gain of 1 means the signal is unchanged in magnitude. In the "stop band" where the signal is being attenuated, the signal's amplitude halves every time the frequency doubles.

The start of the stop band is taken to be after the "corner frequency" (f0), which is defined as the frequency at which the signal amplitude has been reduced to 75% for its original value.

The corner frequency is given by the formula:

$$f_c = \frac{1}{2\pi RC}$$

Using a resistor of 270Ω and 330nF capacitor the corner frequency is:

$$f_c = \frac{1}{2\pi RC} = \frac{1}{2\pi \times 270 \times 330n} = 1.786 kHz$$

With a corner frequency of 1.768kHz and a frequency of 32.7kHz for the PWM signal (that we are trying to filter out) a quick check will show you that you are looking at a little over four doublings of 1.786kHz, which gives a reduction of four halvings of the amplitude of that high-frequency component. This will reduce the amplitude of the 32.7kHz signal to about 1/16 of its unattenuated value, which is why the PWM frequency is still visible on top of the 440Hz sinewave shown in Figure 16-3.

If you want to try out this experiment, then install the Mozzi library into your Arduino IDE by downloading the ZIP for the library from GitHub (*https://github.com/sensorium/Mozzi*) and then installing the ZIP for the library from the Arduino IDE's menu (Sketch→Include Library→Add .ZIP Library).

From the Arduino IDE's example menu, upload Files→Examples→Mozzi→Basics→Sinewave onto your Arduino and wire it up as shown in Figure 16-1.

Attaching an oscilloscope to the unfiltered and filtered outputs of D9 should give you a trace similar to Figure 16-3.

See Also

For a better-quality filter with a much steeper fall-off of gain with frequency, see Recipe 17.7.

In this recipe, we have talked about attenuation of a signal's amplitude in terms of halving or quartering. The more common unit for attenuation and amplification of a signal is the dB (decibel). This is described in Recipe 17.1.

A great way to help design filters is to use simulation. This example is used in Recipe 21.11.

For instructions on using an oscilloscope, see Recipe 21.9.

16.2 Create an Oscillator

Problem

You want to make a simple oscillator using a pair of transistors to flash LEDs or generate an audio signal.

Solution

Use the transistors in the arrangement shown in Figure 16-5. This arrangement is also sometimes called an "astable," meaning it has no stable state, and instead oscillates between two states.

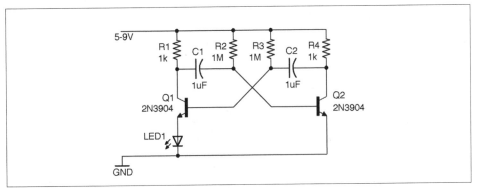

Figure 16-5. A Two-transistor Oscillator

When power is first applied to the circuit both Q1 and Q2 will be off. Very slight differences between the capacitor values and the values of R2 and R3 mean that one of the capacitors will start to charge through R2 or R3 faster than the other. The faster-charging capacitor will win the race to turn on the transistor whose base it is connected to. Current flows through this transistor turning on the LED and drops the collector voltage allowing the slower capacitor to start charging until the opposite transistor turns on and so on.

Discussion

The frequency of this oscillator is determined by the time constant of C1 and R2 (C1 and C2 should have the same value, as should R2 and R3), but also depends on the characteristics of the transistor. For the circuit shown in Figure 16-5 and a supply voltage of 9V, the frequency was measured as 2.8Hz, implying that the frequency is related to R2 and C1 by the formula:

$$f = \frac{1}{0.36R_2C_1}$$

See Also

For an oscillator that uses the NE555 timer IC, see Recipe 16.5.

You can see a video of this circuit in action here: *https://youtu.be/-NvMFmPHc4s*.

16.3 Flash LEDs in Series

Problem

You want to make an odd number (three or more) of LEDs flash in series without using a microcontroller or digital IC.

Solution

Build a ring oscillator using MOSFETs as shown in the schematic in Figure 16-6.

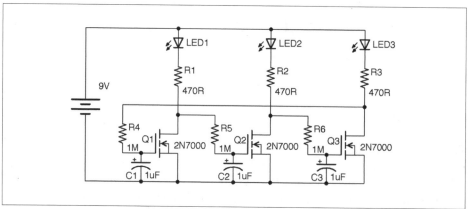

Figure 16-6. A Ring Oscillator

The way to understand how this circuit works is to think of each transistor stage as an inverter. That is, if the gate of the transistor is low, then the drain of the transistor will be pulled high by the LED and resistor. As the transistor's associated gate capacitor charges, the gate voltage will gradually rise until the transistor is on and then the drain will go low. The final inverter stage then feeds back to the first transistor and the cycle continues. This does not work with an even number of stages as the outputs of the first and last stages would both be high and low (in phase) with each other and no oscillation would occur.

Discussion

This oscillator produces quite a mesmeric effect as the LED brightness gradually increases and decreases.

In practical terms, when considering a large number of LEDs, you would probably use a microcontroller with LEDs attached to digital outputs, or a decade counter like the 4017 described in Recipe 15.6.

See Also

For an interesting Wikipedia article on the ring oscillator, see *https://en.wikipedia.org/wiki/Ring_oscillator*.

There is a video of this circuit here: *https://youtu.be/9O5Ojhr0oGg*.

16.4 Avoid Drops in Voltage from Input to Output

Problem

You need to buffer a high-impedance input voltage so that it can be attached to a significant load at its output without affecting the input.

Solution

Use a BJT arranged as an emitter-follower, as shown in Figure 15-7. In this case, the input is provided by a pot. The output voltage will follow the input voltage (less the base-emitter voltage).

On the face of it, this may seem pointless and you might be asking yourself why you wouldn't just take the output from the slider of R1. The reason is that even a very slight load resistance on R1 will alter its output voltage, whereas by using a transistor, a much larger load current can be drawn.

Whatever the voltage set at the base of the transistor (above about 0.6V), the emitter will always be about 0.6V lower, but able to provide significantly more current. In fact, the ratio of base-to-collector current is the DC gain of the transistor. For a transistor like the 2N3904, this is generally taken to be about 100. So for a collector current of say 10mA, less than 100µA should be flowing into the base from R1.

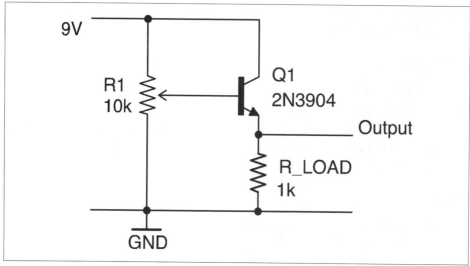

Figure 16-7. An Emitter-Follower Used as a Buffer

Discussion

An emitter-follower like this can be used as the basis of a voltage regulator using the schematic in Figure 16-8, although generally a voltage-regulator IC is an easier option.

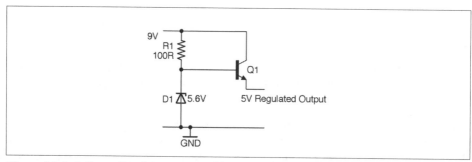

Figure 16-8. A Zener Diode/Emitter-Follower Voltage Regulator

As described in Recipe 4.3 the Zener diode will keep the base of Q1 at 5.6V as long as the input voltage is greater than 5.6V. The Zener diode acts as a voltage reference to Q1, whose emitter will remain at 5V as long as the load current does not become great enough to start drawing significant base current and hence cause the voltage at the base to drop. Using a power Darlington like the TIP120 with its DC current gain of around 10,000 will provide even better regulation, at higher currents, but at the expense of a bigger base-emitter voltage drop (see Recipe 5.2).

See Also

For a more or less perfect unity gain buffer without the base-emitter voltage drop, see Recipe 17.6.

For more information on BJTs, see Recipe 5.1.

16.5 Build a Low-Cost Oscillator

Problem

You want a simple low-cost oscillator (a.k.a. astable) that produces a 50% duty cycle and has a push-pull output capable of 200mA.

Solution

Use an NE555 timer IC in the configuration shown in Figure 16-9.

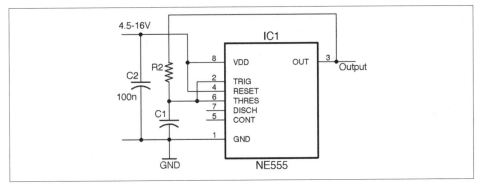

Figure 16-9. An NE555 Oscillator

This is not the standard astable oscillator schematic for an NE555. But unless you need to control the duty cycle for PWM or other purposes, it's fine for a simple oscillator.

The frequency is set by the values of R2 and C1:

$$f = \frac{0.693}{R_2 C_1}$$

So, for an R2 of 10kΩ and a C1 of 10nF, the frequency is:

$$f = \frac{0.693}{R_2 C_1} = \frac{0.693}{10k \times 10n} = \frac{693}{100} kHz = 6.93 kHz$$

To simplify the calculations, standardize a few values of C1. For low-frequency LED blinking, use a 1µF capacitor; for audio frequencies of a few hundred Hz, use 100nF; and for higher frequencies into the kHz, use a 10nF capacitor. Fixing the capacitor in this way means you can use the following formula to calculate the value of R2 that you need:

$$R_2 = \frac{0.693}{fC_1}$$

Table 16-1 shows some common frequencies and suitable component values.

Table 16-1. Values for Common Components

Frequency	C1	R2
1Hz	1µF	693kΩ
2Hz	1µF	347kΩ
50Hz	1µF	13.9kΩ
100Hz	1µF	6.93kΩ
1kHz	10nF	69.3kΩ
10kHz	10nF	6.93kΩ
100kHz	10nF	693Ω

Discussion

Remember that capacitors in particular are generally only accurate at values of ±10% and the frequency is also somewhat dependent on the supply voltage. Table 16-2 shows the effect of supply voltage on output frequency for this circuit using an R2 of 10kΩ and a C1 of 10nF.

Table 16-2. Sensitivity of Output Frequency to Supply Voltage

Supply Voltage (V)	Output Frequency (kHz)
5	5.46
9	6.63
12	7.03
16	7.33

See Also

You can find the datasheet for the NE555 here: *http://www.ti.com/lit/ds/symlink/ne555.pdf*.

If you need complimentary outputs from your oscillator, use a 4047 timer IC (Recipe 7.10).

The NE555 IC is a versatile device and can also be used as a one-shot timer (Recipe 16.7).

You can also make an oscillator with just two transistors as described in Recipe 16.2.

16.6 Build a Variable Duty Cycle Oscillator

Problem

You need an oscillator (a.k.a. astable) but want to set the duty cycle.

Solution

Use an NE555 timer IC, but this time in the configuration shown in Figure 16-10.

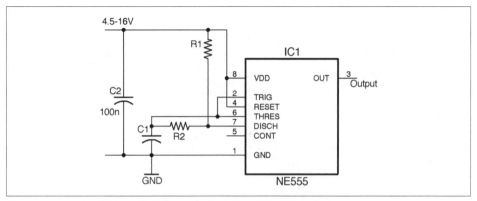

Figure 16-10. An NE555 Oscillator with Adjustable Duty Cycle

The frequency and duty cycle (proportion of the time the output is high to the time it is low) are set by the values of C1, R1, and R2. The period of time the output is high during each cycle is:

$$T_{high} = 0.693(R_1 + R_2)C_1$$

The time it is low is given by:

$$T_{low} = 0.693 \times R_2 C_1$$

The overall frequency in Hz will be the inverse of these two times added together:

$$f = \frac{1}{T_{high} + T_{low}} = \frac{1.44}{(R_1 + 2R_2)C_1}$$

If you want a fairly even duty cycle, you can just use Recipe 16.5, or make sure that R1 is much smaller than R2 (without being 0).

To avoid all the math, there are lots of good online calculators out there you can use. This one (*http://www.daycounter.com/Calculators/NE555-Calculator.phtml*) is particularly useful as it allows you to enter just the frequency and duty cycle you want.

Discussion

The NE555 timer is an extremely flexible device. Figure 16-11 shows the internal structure of the IC.

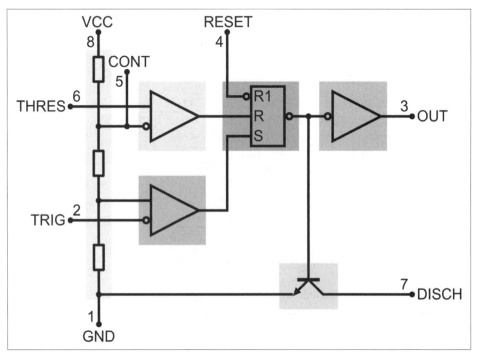

Figure 16-11. A Look Inside an NE555 Timer IC

The design is centered around a reset-set (RS) flip-flop (see Recipe 15.4). This flip-flop has an output that will become high if the S input goes high. It will then stay high until the flip-flop is reset, either by R becoming high or R1 becoming low. The output of the flip-flop drives both a push-pull output stage (Recipe 11.8) connected to the OUT pin and an open-collector output connected to pin DISCH, which is used to discharge a timing capacitor.

Setting and resetting the flip-flop is achieved by using an arrangement of three resistors in a voltage divider from the positive supply (VCC) to GND and two comparitors (Recipe 17.10). If the voltage at TRIG falls below ⅓ of the supply voltage, the lower comparitor sets the flip-flop until the voltage at THRES exceeds ⅔ of the supply voltage and the flip-flop is reset.

The CONT (control) pin is not generally used, but it can be used to adjust the comparitor-threshold voltages. You will also often find schematics where a 10nF capacitor is connected between CONT and GND. This is an additional decoupling capacitor that can improve the oscillator stability, but is by no means essential.

In addition to the standard NE555 timer, a number of variations of the device have been developed. The NE556 is simply a 14-pin IC package that has two NE555 timers in it sharing common supply pins.

The LMC555 is a CMOS version of the 555 timer that is pin compatible and will operate down to 1.5V supply.

See Also

You can find the datasheet for the NE555 here: *http://www.ti.com/lit/ds/symlink/ne555.pdf* and the datasheet for the 555 here: *http://www.ti.com/lit/ds/symlink/lmc555.pdf*.

Figure 16-11 is taken from an informative Wikipedia page on the 555 timer: *https://en.wikipedia.org/wiki/555_timer_IC*.

16.7 Make a One-Shot Timer

Problem

You want to create a timer that at the press of a button will turn on an output for a certain amount of time.

Solution

Figure 16-12 shows an NE555 timer configured as a one-shot (monostable) timer.

When SW1 is pressed, OUT becomes HIGH and stays HIGH until either SW2 is pressed (to cancel the timer) or a period of time 1.1 x R1 x C1 has elapsed.

For example, a value of C1 of 100μF and R1 of 100kΩ will give a delay of:

1.1 x 100μ x 100k = 11 seconds

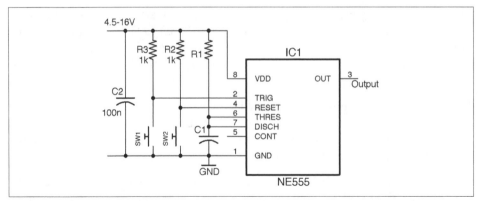

Figure 16-12. Using an NE555 Timer as a One-Shot Timer

Discussion

In this timer circuit to achieve timings of a second or more, you need to use an electrolytic of 100µF or greater.

You can, of course, make R1 a variable resistor, or more likely a fixed resistor and variable resistor in series in order to set a minimum delay.

See Also

For other 555 timer recipes see Recipe 16.5, Recipe 16.6, Recipe 16.8, Recipe 16.9, and Recipe 16.10.

16.8 Control Motor Speed

Problem

You want to be able to control the speed of a motor at the turn of a knob, without having to use an Arduino.

Solution

Use an NE555 timer as shown in Figure 16-13.

For a PWM frequency of around 1kHz, use a value for R1 of 270Ω, R2 a 10kΩ pot, and C1 100nF.

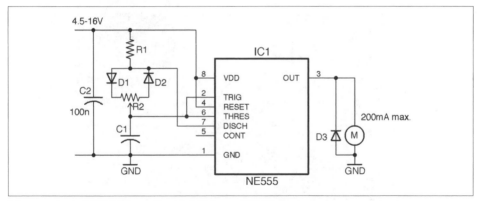

Figure 16-13. PWM Motor Control Using a 555 Timer

Discussion

The NE555 timer output can drive about 200mA, which is only sufficient for a pretty small motor. Drawing more current than this will cause the IC to overheat and eventually fail. To drive a higher power motor, use a transistor as described in Recipe 13.2.

The minimum duty cycle of this circuit depends on the ratio of R1 to R2. R1 should be considerably lower than R2 to minimize the duty cycle, but R1 cannot be 0Ω or the oscillations will stop. A ratio of 40:1 is sufficient to ensure the minimum duty cycle is just 3–4%.

Figure 16-14 shows the minimum duty cycle with the knob of R2 at one end of its travel, Figure 16-15 shows a 50% duty cycle, and Figure 16-16 the maximum duty cycle.

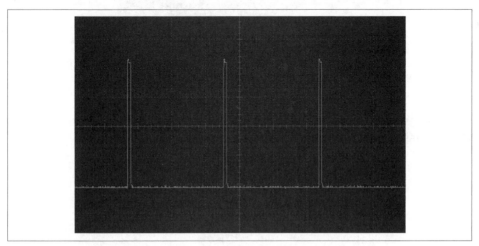

Figure 16-14. PWM Motor Control (Minimum Duty Cycle)

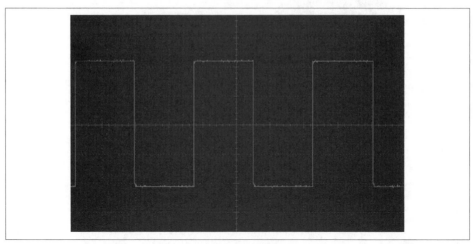

Figure 16-15. PWM Motor Control (50% Duty Cycle)

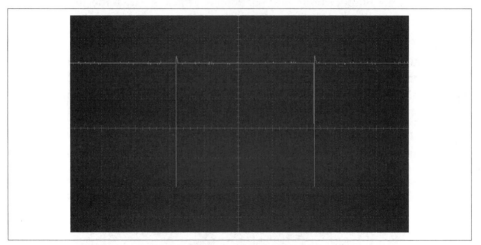

Figure 16-16. PWM Motor Control (Maximum Duty Cycle)

See Also

To use PWM outputs from an Arduino or Raspberry Pi, see Recipe 13.2.

16.9 Apply PWM to an Analog Signal

Problem

You have an analog signal that you want to PWM.

Solution

Use an NE555 timer in the configuration shown in Figure 16-17.

The analog signal must be between 0 and VCC and is applied to the CONT (control) pin of the IC. The Clock signal should be at the modulation frequency and can be provided by another NE555 timer in a stable configuration as described in Recipe 16.5. Rather than using a separate chip to provide the clock, you can use an NE556, which contains the equivalent of two NE555s in one package.

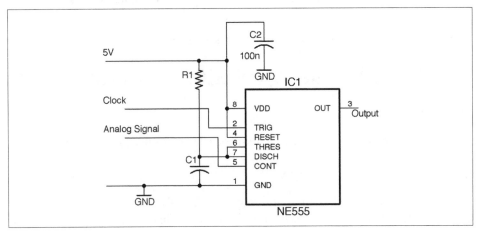

Figure 16-17. PWM of an Analog Signal Using an NE555 Timer

Discussion

The NE555 is configured as a monostable whose pulse length is determined by R1, C1, and the voltage at CONT. This monostable fires every time the TRIG pin pulses in response to the modulation clock.

R1 and C1 set the maximum pulse length, which should be timed to be one wavelength of the clock.

For example, if the clock frequency is 30kHz, the period length is:

$1/30k = 33.3\mu s$

The maximum pulse length T_{max} (see Recipe 16.7) is:

$T_{max} = 1.1\ R1\ C1$

So, an R1 of 470Ω and a C1 of 100nF would give a T_{max} of:

$1.1 \times 270 \times 100n = 29.7\mu s$

This would allow the full voltage range of 0 to 5V to be modulated as PWM.

The use of two NE555 ICs (or a single NE556) configured as an oscillator providing a clock with the second configured as a monostable is an alternative to Recipe 16.8 for using PWM top-control power to a load. In such a case, the CONT pin could be connected to the slider of a pot connected between 5V and GND to control the pulse width.

See Also

Creating a PWM output signal is the first step in class-D digital amplification (see Recipe 18.5).

To control power to say a motor or other load, see Recipe 16.8.

16.10 Make a Voltage-Controlled Oscillator (VCO)

Problem

You want to create an oscillator whose frequency depends on a control voltage.

Solution

Use an NE555 configured as an astable and use the CONT pin voltage to control the switching thresholds and hence the frequency. This arrangement is shown in Figure 16-18.

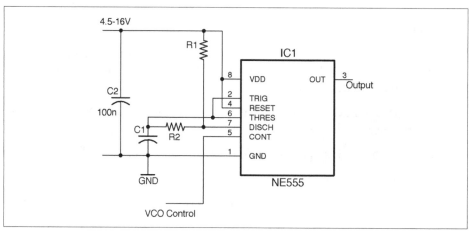

Figure 16-18. Using an NE555 as a VCO

Discussion

VCOs are standard building blocks in analog audio synthesizers, where the output of a low-frequency oscillator can be used to modulate the audio frequency of a VCO.

This VCO design operates over a fairly limited range of frequencies. Using a 5V supply and values of R1 = 1kΩ, R2 = 10kΩ, and C1 = 10nF the plot of Figure 16-19 shows how the frequency varies with the control voltage.

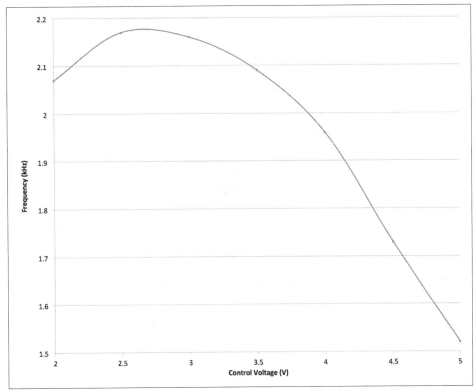

Figure 16-19. Frequency Against Control Voltage

The useful control voltage range is between about 3 and 5V in this example. Oscillation stopped completely below 1.8V.

See Also

For the use of a VCO in an FM transmitter, see Recipe 19.1.

16.11 Explore Decibel Measurement

Problem

You have seen the unit decibel (dB) and want to know what it means.

Solution

The term "decibel" is not strictly speaking a unit of anything, but rather a way of expressing ratios that is suitable for properties such as sound whose perception by humans is logarithmic. That is, for a sound signal to sound a little bit louder, you might need to double the amplitude (voltage).

That is why a 1W amplifier can provide a reasonable volume of sound, but if you need more volume, then rather than take the next step to 2W of output power, you might jump straight to 10W and then on to 100W.

A positive value of dB represents amplification of a signal and a negative dB attenuation (reduction in amplitude) of a signal.

A gain expressed in dB can either apply to voltage (amplitude) or to power, the power being proportional to the amplitude squared. The unit dB is most commonly used with amplitude, in which case the gain is as follows (log is log to the base 10):

$$gain = 20 log \frac{V_{out}}{V_{in}}$$

Figure 16-20 shows various common ratios and the dB equivalents.

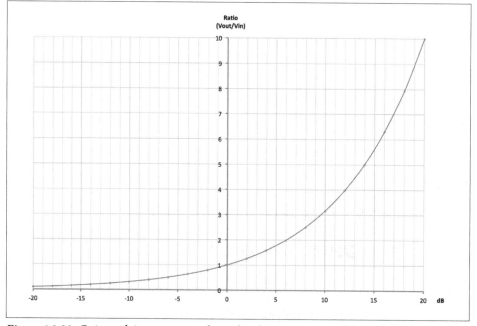

Figure 16-20. Gain and Attenuation of Amplitude Expressed in dB and as a Ratio

A gain of 0dB means the output is exactly the same amplitude as the input. A gain of +6dB means a doubling of the amplitude and –6dB an attenuation of the signal by half.

Table 16-3 shows some common dB ratios and what they mean in terms of amplification or attenuation of both amplitude (voltage) and power.

Table 16-3. dB Amplifications and Attenuations for Amplitude and Power

dB	Amplitude (Voltage)	Power
100	gain of 100,000	gain of 10,000,000,000
80	gain of 10,000	gain of 100,000,000
60	gain of 1,000	gain of 1,000,000
40	gain of 100	gain of 10,000
20	gain of 10	gain of 100
10	gain of 3.162	gain of 10
6	gain of approx. 2	gain of 3.981
3	gain of 1.413	gain of approx. 2
0	no change	no change
−3	attenuation of 1.413	attenuation of approx. 2
−6	attenuation of approx. 2	attenuation of 3.981
−10	attenuation of 3.162	attenuation of 10
−20	attenuation of 10	attenuation of 100
−40	attenuation of 100	attenuation of 10,000
−60	attenuation of 1,000	attenuation of 1,000,000
−80	attenuation of 10,000	attenuation of 100,000,000
−100	attenuation of 100,000	attenuation of 10,000,000,000

Discussion

The term decibel is often used to express the loudness of a sound. Strictly speaking this should be called dBA (decibels Absolute) and relates to the energy in the pressure wave at the point where it is produced, and is an absolute value not a ratio.

See Also

Wikipedia has a useful description of decibels as a unit of gain at *http://bit.ly/2lPeant*.

## CHAPTER 17

Operational Amplifiers

17.0 Introduction

Operational amplifiers, or more commonly known as op-amps, have a helpful way of making theory easy to implement. If you need to make a filter or preamplifier, an op-amp is often the best solution.

An op-amp has two inputs and one output and is represented in schematic diagrams (Figure 17-1) as a triangle with the output coming from one apex and the two inputs labeled + and – on the opposite side. They also need positive and negative power connections. They are available in a variety of IC packages with 5, 6, or 8 pins. ICs are also available that contain two or four op-amps in a single package.

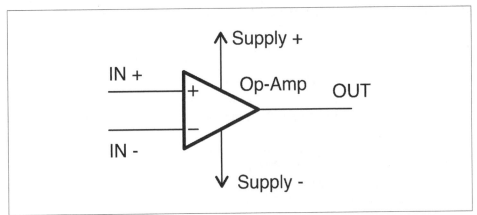

Figure 17-1. Schematic Symbol for an Op-Amp

Op-amps amplify the difference between the + and – inputs. The gain of an amplifier will generally be millions or even billions. So the output voltage may be 1,000,000 times the difference between the input voltages. Such a high gain may sound desirable, but it is actually too high to do anything useful with. To reduce the gain to a more manageable level, an op-amp is almost always used with negative feedback where some of the output is fed back to the negative input. You will find examples of using feedback in this way in Recipe 17.4 and Recipe 17.5. In Recipe 17.6 you will see how if all of the output is fed back to the negative input the output will follow the positive input to the op-amp rather like the emitter-follower described in Recipe 16.4.

Traditionally op-amps are powered with a split power supply that has positive and negative supply voltages (Recipe 17.2) in addition to GND in the middle. However, the prevelance of microcontrollers with single 3.3V or 5V supplies has led to many op-amps being capable of single-supply operation (Recipe 17.3).

An op-amp IC is much better than a discrete-transistor design in most situations, providing a lower component count and generally excellent performance. Special types of op-amps are available for different power supply, frequency, and noise requirements.

17.1 Select an Op-Amp

Problem

Look at any online component catalog and you will find literally thousands of models of op-amps. You want to select the right one for your project.

Solution

Here I have tried to simplify this enormous choice to a small set of readily available devices to suit all applications.

Factors that you need to consider include:

- Price
- Power supply range
- Whether you need the outputs and inputs to be able to reach the full supply range (rail to rail)
- Speed. The gain bandwidth product is the frequency at which the gain of the op-amp (without any feedback) will drop to 1.
- Output slew rate. The maximum rate at which the output can increase.
- Common mode rejection (CMR). A op-amp amplifies the difference between its two inputs, and the CMR of an op-amp is the degree to which any simultaneous changes to both inputs are ignored. This is measured in decibels (see Recipe 16.11).

- Noise. All electronic circuits generate a small amount of noise. This can cause problems with very weak signals, so sometimes in high gain applications low-noise op-amps are necessary.
- Supply current. Some op-amps operate at such low currents that they are suitable for long-term battery use. Some op-amps also have a power-down or standby mode that allows them to be virtually switched off. For example, a microcontroller might do this before putting itself to sleep to save battery power.
- Output current. Sometimes a high output current from an op-amp can be useful to directly drive a small load.
- Number of op-amps per package. Op-amps are available in packages that contain 1 to 4 op-amps in a single package. If you have a circuit that uses multiple op-amps, this can save space and money in your design.

A good selection of devices to start from are listed in Table 17-1.

Table 17-1. Selecting the Right Op-Amp

	LM741	LM321	TLV2770	OPA365
Description	Perhaps the longest-lived and most popular op-amp.	A good high-supply voltage device capable of single-supply operation.	A single-supply low-voltage device with a standby, low-current mode.	A high-performance, low-voltage, low-noise device. SMD only.
Guide cost	$0.50	$0.70	$2	$2
Supply voltage range	±10–22V	3–30V	2.5–5.5V	2.2–5.5V
Rail to rail	N	N	Y	Y
Gain bandwidth product	1MHz	1MHz	5.1MHz	50MHz
Slew rate	0.5V/µs	0.4V/µs	10V/µs	25V/µs
Common mode rejection	96dB	85dB	86dB	120dB
Noise (nV/√Hz)	Not specified	40	17	4.5
Supply current	1.7mA	0.7mA	1mA (1µA standby)	4.6mA
Output current	25mA	20mA	50mA	65mA
Dual-amp package	LM747	LM358	TLV2773	OPA2385
Quad-amp package	LM148	LM324	TLV2775	Not available

Discussion

There will be times when you need to step out from the selections in Table 17-1 for some particular application. Perhaps you need an op-amp that works at high frequencies, but has a high upper supply voltage. In such cases, there will almost certainly be a device that fits the niche, but to find it, you will need to do some searching on the internet for recommendations and then carefully inspect the datasheets.

Note that not all op-amps have the same pinout. See Appendix A for pinouts for the op-amps described in this recipe.

See Also

Take a look at the datasheets for the op-amps featured in this recipe:

- 741 (*http://www.ti.com/lit/ds/symlink/lm741.pdf*)
- LM321 (*http://www.ti.com/lit/ds/symlink/lm321.pdf*)
- TLV2770 (*http://www.farnell.com/datasheets/1962331.pdf*)
- OPA365 (*http://www.ti.com/lit/ds/symlink/opa2365.pdf*)

You will also find an op-amp used in Recipe 18.3.

17.2 Power an Op-Amp (Split Supply)

Problem

You need a power supply suitable for powering an op-amp such as the LM741, which requires a split-supply with positive, negative, and ground.

Solution

Use a regulated supply or batteries, or both. In any case, use a linear voltage regulator rather than a switching regulator. The current consumption of an op-amp is low enough that power-supply efficiency does not need to be considered.

If you are operating the op-amp with a split power supply, you can use positive and negative versions of a linear regulator as shown in the ±12V supply shown in Figure 17-2.

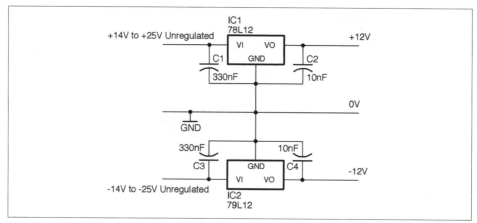

Figure 17-2. A Split 12V Regulated Supply

Op-amps benefit greatly from the use of decoupling capacitors (see Recipe 15.1). If you have an undemanding application, a single 100nF capacitor placed close to the op-amp IC will be just fine. For more demanding applications where you have a lot of gain, it is common to use both a 100nF and 10μF capacitor next to each other and close to the IC body.

Discussion

The positive side of the supply is as described in Recipe 7.4, but in this case, in addition to a 78L12 (L for low-power) to regulate the positive side of the supply, a 79L12 negative voltage regulator is used to regulate the negative side of the supply.

See Also

Although a split supply like this is quite common in audio and scientific instruments, in general, it is more convenient to use a single supply as described in Recipe 17.3.

17.3 Power an Op-Amp (Single Supply)

Problem

You want to use an op-amp with a single-supply voltage, but also need to provide a reference voltage midway between the supply rails.

Solution

Figure 17-3 shows a common way of providing a 5V regulated supply and a 2.5V reference voltage using a voltage divider and capacitor. The capacitor C3 stabilizes the voltage further, making it less likely to be influenced by changes in the current flowing through the reference voltage.

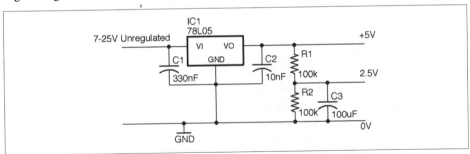

Figure 17-3. Providing a Center Voltage for Single-Supply Operation of an Op-Amp

As you will see in Recipe 17.4 and Recipe 17.5 nearly all op-amp circuits need this middle reference voltage to provide negative feedback to reduce the op-amp's gain.

Discussion

To improve this and provide a stable center voltage use a voltage divider and capacitor with a unity gain buffer as described in Recipe 17.6. If you have a design that uses several op-amps, you will often find that there is a "spare" op-amp in a quad op-amp package that can be used for this purpose.

See Also

To power an op-amp using a split-voltage supply, see Recipe 17.2.

17.4 Make an Inverting Amplifier

Problem

You need a simple op-amp amplifier and it does not matter that the signal output will be the inverse of the input.

Solution

Connect up your op-amp as shown in the schematic in Figure 17-4, which shows the arrangement assuming a split ±12V supply.

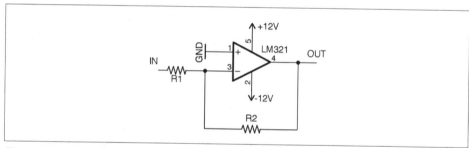

Figure 17-4. An Inverting Amplifier

Without any feedback from the output of an op-amp to its input, the gain of the op-amp will be millions or billions, which is far too high to do anything useful other than amplify noise. This circuit uses R1 and R2 to reduce the gain of the op-amp to a reasonable level. A reasonable level is a factor between 10 and 10,000. If you need more gain than this, you will need to use multiple stages of amplification as well as some filtering to avoid amplifying unwanted noise.

The voltage gain of the circuit is set by the formula:

$$gain = \frac{V_{out}}{V_{in}} = -\frac{R_2}{R_1}$$

So if we want our amplifier to have a gain of −10 (times 10 but inverted), we could choose a value for R2 of 10kΩ and of R1 of 1kΩ. That way when we put +1V at IN, OUT would have a voltage of −10V. Conversely, if we put −0.1V at IN, OUT would be +1V.

If you want to use a single 5V supply, the same principle applies, but now the amplification takes place relative to the center voltage of 2.5V. Figure 17-5 shows the schematic for using a 5V regulated supply with an OPA365 rail-to-rail single-supply op-amp.

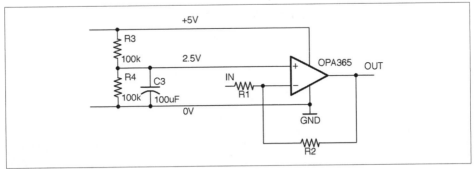

Figure 17-5. Single-supply Inverting Amplification with an Op-Amp

With single-supply operation, the voltage at IN must be between 0 and 5V and the amplification is relative to the center voltage of 2.5V. This means that if the voltage at IN is 2.5V then OUT will also be 2.5V, but if IN is 2.6V (0.1V higher than 2.5V) then OUT will be at 1.5V (1V less than 2.5V).

So, the formula to calculate Vout for any given Vin in the schematic in Figure 17-5 is:

$$V_{out} = (2.5 - V_{in})\frac{R_2}{R_1} + 2.5$$

Discussion

No matter what the gain of your amplifier is, the op-amp's output cannot exceed that of the supply voltages. A rail-to-rail op-amp will allow the voltage to swing from one supply to another but not beyond, but a non–rail-to-rail type op-amp may only allow you to get within a volt or two of the supply voltages.

See Also

Recipe 17.5 shows a noninverting op-amp configuration.

17.5 Make a Noninverting Amplifier

Problem

You want to amplify the voltage of a signal, without inverting it.

Solution

Use an op-amp in the noninverting configuration of Figure 17-6.

In this case, the gain of the amplifier is given by the formula:

$$gain = \frac{V_{out}}{V_{in}} = 1 + \frac{R_2}{R_1}$$

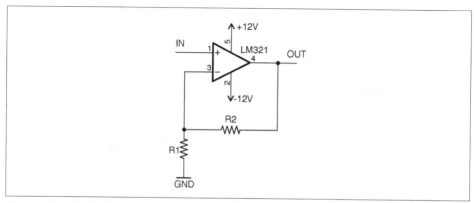

Figure 17-6. A Noninverting Op-Amp Amplifier

If we used a 10kΩ resistor as R2 and a 1kΩ resistor as R1, the gain would be 11. In other words, the voltage at OUT would be 11 times the voltage at IN. If the voltage at IN is negative, the voltage at OUT will also be negative.

In the case of single-supply operation of the amplifier, let's say 5V, we have the same 2.5V offset to contend with as we did in Recipe 17.4. Figure 17-7 shows how you might build a single-supply noninverting amplifier.

In this case, the gain is still:

$$gain = \frac{V_{out}}{V_{in}} = 1 + \frac{R_2}{R_1}$$

And the output voltage is related to the input voltage by the equation:

$$V_{out} = (V_{in} - 2.5)(1 + \frac{R_2}{R_1}) + 2.5$$

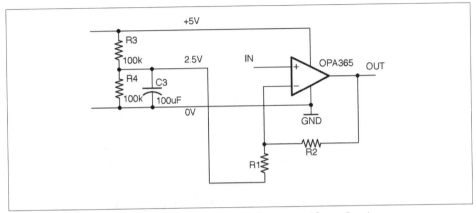

Figure 17-7. Single-supply Noninverting Amplification with an Op-Amp

Discussion

The amplifiers described thus far are capable of amplifying both a DC signal (say from a sensor) and AC-changing signals (perhaps an audio signal). A single-supply 5V (or even 3.3V) design like this is ideally suited to connecting the output to the analog input of a microcontroller, as the voltage can swing between 0V and the supply voltage. In such cases, it is conceptually easier to deal with amplification of a voltage without the added complication of the amplified voltage being inverted about 2.5V, so it is more common to use this noninverting recipe in single-supply designs.

See Also

See Recipe 17.4 for inverting amplifier designs.

17.6 Buffer a Signal

Problem

You need to buffer a signal; that is, you don't want to amplify or attenuate the signal at all, but effectively give it a low output impedance so that it is not affected significantly by loading.

Solution

Configure the op-amp as shown in Figure 17-8.

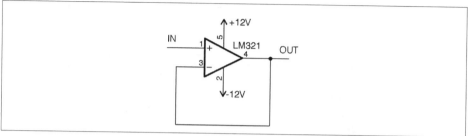

Figure 17-8. A Unity Gain Buffer

The output voltage will track the input voltage. So if the input is 1V, the output will also be 1V. You can then hang whatever load you like off the output, without worrying too much about it affecting the input.

In other words, the voltage is not being amplified but the current that can pass through a load is greatly increased. An example where this might be used is as a headphone amplifier where the signal voltage is high enough in theory to drive the headphones, but the output would not otherwise be sufficient to power the low impedance of the headphones.

Discussion

One situation where a buffer like this can be used is to provide the center voltage for a single-supply op-amp amplifier. Figure 17-9 shows a 5V single supply amplifier modified from Figure 17-7 to use a buffer in this way.

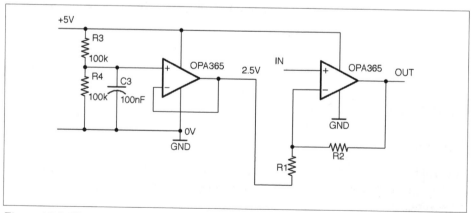

Figure 17-9. Using a Unity Gain Buffer to Provide the Center Voltage for a Single-Supply Noninverting Amplifier

Because the input impedance to the buffer is very high, the value of C3 can be reduced as very little current at all will be flowing out of it. Think of it as reducing ripple on an unloaded power supply.

See Also

Another way to think of the action of a unity gain buffer is as a current amplifier, which is like using a bipolar transistor in an emitter-follower configuration in Recipe 16.4.

17.7 Reduce the Amplitude of High Frequencies

Problem

You want a low-pass filter to reduce the amplitude of high frequencies, but you need it to have a steeper cutoff than the simple RC filter of Recipe 16.3.

Solution

Use an op-amp and a 2-pole filter.

As an example, let's say we want to revisit the crude filter you saw in Recipe 16.3 designed to separate a 440Hz signal from a 32.7kHz PWM carrier frequency. In that simple RC filter, we ended up with a corner frequency of 1.786kHz and saw that there would be a halving of signal amplitude for each doubling of the frequency. This resulted in a reduction in the amplitude of about a factor of 16 at 32kHz. By using an active filter, the attenuation (fall-off) at 32kHz can be improved to a factor of 100 easily.

When selecting component values for R1, R2, and C2, you can do a whole load of complicated math by hand, or you can use a filter design tool. In this recipe, we will use the Analog Filter Wizard provided by the chip manufacturer Analog Devices available as an online tool at *http://www.analog.com/designtools/en/filterwizard/*.

When you navigate to that page in your browser, you will be offered the choice of low-pass, high-pass, or band-pass. When you select low-pass, you will be given the opportunity to specify the corner frequency and other parameters for the filter as shown in Figure 17-10.

Logarithmic Scales

At first sight, the frequency scale of Figure 17-10 looks a little strange as the frequencies 1kHz, 10kHz, and 100kHz are each the same distance apart. That is, the distance along the axis from 1kHz to 10kHz is the same as the distance from the much bigger range of frequencies from 10kHz to 100kHz. Also, the vertical lines are not evenly

---

spaced, but far apart at first, getting closer as they approach the next marked frequency.

This is because in Figure 17-10 and for that matter in Figure 17-12 and Figure 17-15 the frequency scales are logarithmic. That is, to be able to see the full shape of the curve for such a wide range of frequencies, the scale is distorted in length based on the log (base 10) of the frequency.

Looking at the area of the scale around 1kHz, the next vertical line is 2kHz and the one after that 3kHz, gradually getting closer together as they approach 10kHz.

If you tried to plot Figure 17-10 without using a logarithmic frequency scale, the plot would just look like the edge of a cliff. You would not be able to see the details of the frequency response as well as with a log plot.

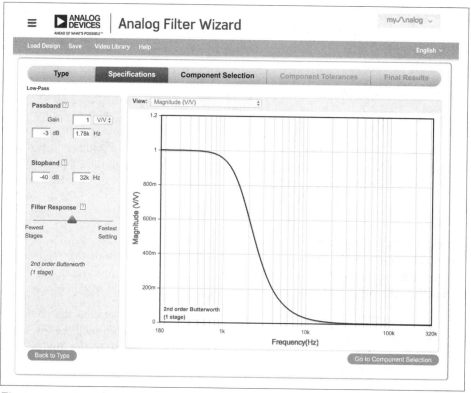

Figure 17-10. Specifying a Filter in the Analog Filter Wizard

The tool is by default configured to work in units of dB (see Recipe 16.11). You can change this to just the amplitude voltage as a ratio by changing the drop-down lists under Passband and next to View to both be V/V as shown in Figure 17-10, where the corner frequency has been set to 1.78kHz and the desired stop-band set to –40dB (factor of 1/100) at 32kHz.

Use the filter-response slider to determine just how steep the curve will be as well as how flat the pass-band's response is. This will also alter the number of op-amps needed for the design (stages) and the order of the filter. A single op-amp can implement a second-order filter. More orders than that and you will need to chain together several op-amps. The filter type (Butterworth, Chebyshec, or Bessel) will all use the same schematic but different component values to provide different behaviors. See the discussion section for the differences.

For the sake of this example, position the slider so that the second-order Butterworth (first stage) is recommended by the tool and then click Go to Component Selection and "presto," a schematic complete with component values will be recommended to you as shown in Figure 17-11.

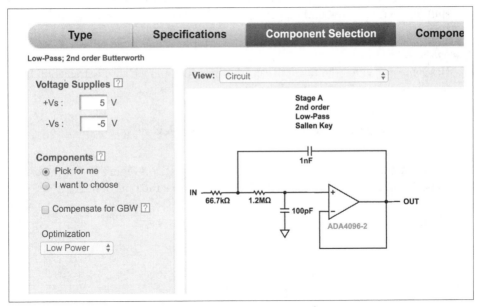

Figure 17-11. Filter Design Complete with Component Values

You can then specify supply voltages and other properties of your design as well as look at the effect of component accuracy on the design. Since this is an Analog Devices tool, it will obviously recommend the op-amps they manufacture, but the designs can be used with other manufacturers' op-amps.

There is nothing to stop you from adding some gain to your filter so that it both filters and amplifies by replacing the connection from the op-amp output to the negative input with a pair of resistors as used in the noninverting amplifier design of Recipe 17.5. In fact, the design tool does allow you to specify a gain when you are specifying the filter (see Figure 17-10).

Discussion

Unless you have a particularly demanding project, a second-order filter using a single op-amp will generally provide a good balance between complexity and performance. There are whole books written on filter design and other more complex filter design tools.

The properties of the three filter types offered by the Analog Filter Wizard are:

Butterworth
  A very flat response (the gain remains the same) for the pass-band before you get to the corner frequency, but the transition from pass-band to stop-band is not as steep as for other filter types. Butterworth filters introduce a frequency-dependent phase-shift between the input and output signals, which leads to distortion of the original signal.

Chebyshev
  Has some gain ripples in the pass-band, but provides a much steeper transition from pass-band to stop-band, although it also suffers from phase-shift.

Bessel
  Has very little phase-shift at the expense of an even less steep transition from pass-band to stop-band.

These types are all implemented using the same schematic in Figure 17-11; the calculations performed by the Analog Filter Wizard determine the type of filter.

See Also

For other types of filter designs, see the next few recipes.

17.8 Filter Out Low Frequencies

Problem

You want to make an active filter that will discard low frequencies and leave higher frequencies unaltered.

Solution

Use a second-order filter implemented using a single op-amp and design it using a design tool such as the Analog Filter Wizard. Figure 17-12 shows the frequency response of a filter designed by the Analog Filter Wizard to provide high-pass filtering with a gain of 10 and a corner frequency of 1kHz.

Notice that there is a little "hump" in the response between 2kHz and 3kHz where the gain goes above the desired value of 10. This is called *overshoot* and is another characteristic of Chebyshev filters.

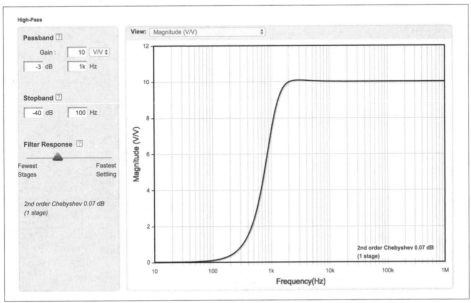

Figure 17-12. Frequency Response of a Second-Order Chebyshev High-pass Filter

The corresponding schematic for this filter is shown in Figure 17-13.

Discussion

Most of the things said in the discussion section of Recipe 17.7, especially about the different types of filters, apply equally well to high-pass filters.

Notice that in Figure 17-13 the Analog Filter Wizard has also helpfully suggested the necessary bypass capacitor values (Recipe 15.1).

See Also

For a low-pass filter see Recipe 17.7 and for a band-pass filter see Recipe 17.9.

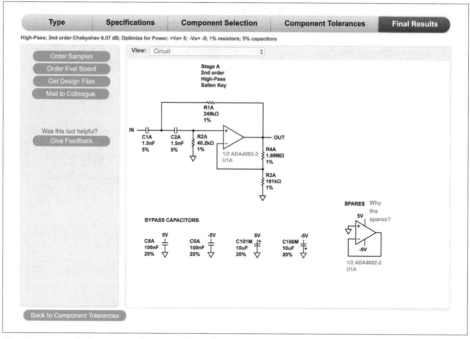

Figure 17-13. Schematic of a Second-Order Chebyshev High-pass Filter

17.9 Filter Out High and Low Frequencies

Problem

You want a filter that attenuates frequencies outside of the range of frequencies that you want to keep.

Solution

If the band of frequencies you are interested in is reasonably wide, then the best solution is to simply pass the signal through a low-pass filter (Recipe 17.7) and then a high-pass filter (Recipe 17.8.) This will require two op-amps, but then dual op-amp ICs are not much more expensive than single op-amp ICs.

Figure 17-14 shows the schematic for a band-pass filter with corner frequencies of 20Hz and 20kHz, which might be used to restrict a signal to the audio-frequency range.

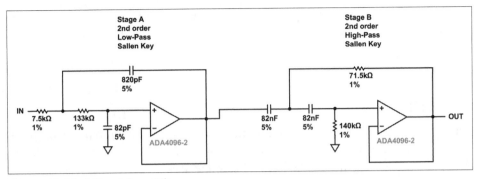

Figure 17-14. A Band-Pass Filter

The filter was designed using the Analog Filter Wizard used in Recipe 17.7 and Recipe 17.8. This time the filter type of Band-pass is selected and the parameters entered as shown in Figure 17-15.

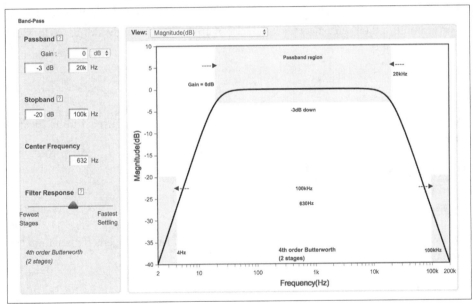

Figure 17-15. Designing a Band-pass Filter Using the Analog Filter Wizard

The pass-band is set to 20kHz for a 3dB drop in amplitude (0.71 of the original amplitude) and the stop-band to 100kHz at the high end where we are aiming for attenuation of 20dB (0.1 of original signal amplitude).

The center frequency is not the 10kHz that you might expect but rather the square root of the product of the upper- and lower-corner frequencies. In other words:

$$\sqrt{20 \times 20,000} = 632Hz$$

Discussion

Filter design is a specialized field, and in this book, there is only room to look at the most common electronics recipes. However, for many applications a simple design that uses a single op-amp stage will be perfectly adequate.

Q-Factor

The Q or "quality" factor of a band-pass filter is an indication of the narrowness of the band and is the ratio of the center frequency of the band to its bandwidth. The filter's bandwidth is the band that is passed without attenuation of more than 3dB.

For example, a filter with a center frequency of 10kHz and a bandwidth of 5kHz will have a Q factor of 2.

See Also

For low-pass active filter design see Recipe 17.7 and for high-pass filter design see Recipe 17.8.

17.10 Compare Two Voltages

Problem

You want to compare two voltages and switch a load if one voltage is higher than the reference voltage.

Solution

Use a device closely related to the op-amp called a *comparator*. Figure 17-16 shows how you could make a simple automatic light that turns on an LED when the light level falls below a certain threshold.

R1 and R3 form a voltage divider whose output corresponds to the light level; that is, the higher the volate, the higher the light level. The variable resistor R2 is used to set the reference voltage at IN–.

The output of the LM311 provides optimal flexibility by providing connections to both the emitter and collector of the NPN output transistor.

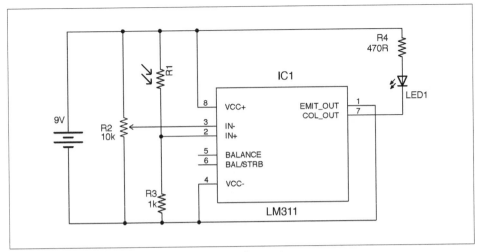

Figure 17-16. Using a Comparator to Control an LED

If the light level is low enough for IN+ to be lower than IN– the base of the output transistor is supplied with current, and since the emitter is grounded, the transistor turns on and current flows through the LED.

This logic of the transistor being on if IN+ is lower than IN– would seem to be counterintuitive, but the output transistor will often have its emitter grounded and the logic output taken from the collector, which is pulled up to VCC. This effectively acts as an inverter and hence the collector output will be high if IN+ is greater than IN–.

Discussion

You can use an op-amp as a comparator, but it is better to use a special-purpose comparator IC because it will generally have a higher output current drive and voltage range.

See Also

To sense light using an Arduino, see Recipe 12.3 and for a Raspberry Pi Recipe 12.6.

## CHAPTER 18

Audio

18.0 Introduction

Audio electronics is all about generating signals and amplifying them enough to power a loudspeaker so that you can hear them.

This chapter contains a number of amplification and signal-generation recipes. Any resource that deals with audio-power amplifier design will explain the different classes of amplifiers, which are designated by A, B, AB, and D, and some other more exotic designs. For the most part, these letters designate how the amplifier is constructed from bipolar junction transistors and how they are biased. Since these days there is no real cost or even quality advantage (apart from the most demanding audiophile) to building an amplifier from separate transistors rather than just using an IC, I feel it is only necessary here to highlight the most common designs along with their main characteristics:

Class A
> Low distortion but very inefficient, produces a lot of heat.

Class B and AB
> A push-pull design. More distortion than A but much more efficient. AB is the same as B but with compensation to reduce the distortion when the amplifier switches over from pushing the loudspeaker to pulling it (see Recipe 18.4)

Class D
> Digital amplifier. Higher distortion than A or B but very energy efficient (see Recipe 18.5)

Before launching into the construction of your own amplifier designs, think about whether you need to build the amplifier from scratch. For most one-off projects, it is easier and cheaper to use ready-made modules or even USB-powered amplified speakers.

When it comes to generating sound waves from an Arduino, you can easily generate tones. The Raspberry Pi with its large amount of memory and sophisticated hardware allows MP3 and other sounds files to be played through its audio jack, which can then be connected to an external amplifier.

18.1 Play Sounds on an Arduino

Problem

You want to know how to generate sounds from an Arduino and play them through a loudspeaker.

Solution

You can experiment with generating tones and driving a speaker from an Arduino if you connect a speaker to an Arduino as shown in Figure 18-1.

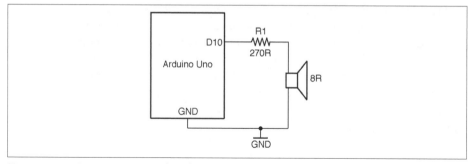

Figure 18-1. Generating Sound with an Arduino

The sketch *ch_18_speaker* can be found with the downloads for the book (see Recipe 10.2):

```
const int outputPin = 10;

void setup()
{
  pinMode(outputPin, OUTPUT);
  Serial.begin(9600);
  Serial.println("Enter frequency 100-8000 Hz (0 off)");
}

void loop()
```

```
{
  if (Serial.available())
  {
    int f = Serial.parseInt();
    if (f == 0) {
      noTone(outputPin);
    }
    else {
      tone(outputPin, f);
    }
  }
}
```

The Arduino tone function takes two parameters, the pin to generate a tone on and the frequency of the tone to generate as a long integer value. The tone function can generate frequencies from 31Hz to 65,535Hz.

To use it, open the Arduino Serial Monitor and make sure you have the Line Ending drop-down list set to "No line ending" then type in a frequency in Hz and click Send.

The 5V signal generated by the tone command is a little too high a voltage to feed directly into an amplifier. It should be reduced to below 2V or so using a pair of resistors as a voltage divider. 1kΩ and 270Ω resistors with the 270Ω resistor to GND will reduce the output by a factor of about 5.

Loudspeakers

Figure 18-2 shows the structure of a loudspeaker.

There are two main parts of a loudspeaker: the frame that holds everything in place and the cone that moves back and forth to create the pressure waves in the air that we hear as sounds.

The frame includes a fixed magnet and the cone, which has a coil of wire wound around the narrow end. Flexible wires link the coil to terminals on the frame. When a current passes through the coil it causes movement in the cone relative to the fixed magnet, producing pressure waves in the air.

Speakers that produce a reasonable amount of sound are quite high power, often tens of watts. In order to avoid the dangers of having to use high voltages to drive loudspeakers, they are usually low resistance. 8Ω is a common value and even 4Ω speakers are quite common, especially in cars where only 12V DC is available.

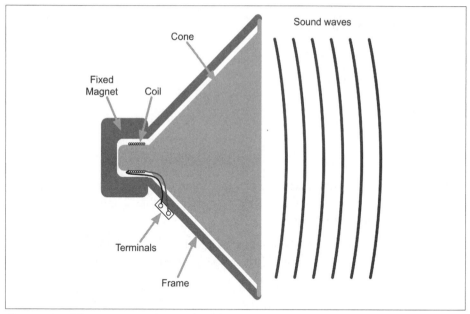

Figure 18-2. The Structure of a Loudspeaker

Discussion

A loudspeaker is an inductive load, so, like the coil of a relay, there is a risk of voltage spikes arising that could damage the GPIO pin driving the speaker. In reality, when using a small speaker driven through a resistor, these spikes will be sufficiently low in energy for the antistatic protection of the GPIO pin to take care of them.

Making crude square-wave buzzing has its place, but the Arduino is actually capable of more sophisticated sound generation using the Mozzi library, which uses a high-frequency PWM signal at 32.7kHz to generate the sound file using PWM.

To try out Mozzi and hear the difference between the square and sinewaves, install the Mozzi library (*http://sensorium.github.io/Mozzi/*) and then run the example program included with the Mozzi library 01. Basics→Sinewave.

You will need to change the connection to R1 from Arduino pin 10 to pin 9 that is used by Mozzi.

If you want to know how Mozzi sounds are generated skip ahead to Recipe 18.1.

See Also

For using a voltage divider, see Recipe 2.6.

For oscillator recipes that do not require an Arduino, see Recipe 16.2, Recipe 16.5, and Recipe 16.6.

To generate better-sounding audio from an Arduino, take a look at the Arduino Mozzi library (*http://sensorium.github.io/Mozzi/*).

18.2 Play Sound with a Raspberry Pi

Problem

You want to play a sound using the Raspberry Pi's audio jack.

Solution

When it comes to hardware, most models of Raspberry Pi have a 3.5mm headphone-style jack socket that can be attached to a stereo amplifier and speakers.

In recent versions of Raspbian, you will find the OMXPlayer preinstalled. To play a sound file, you just need to name the sound file you want to play OMXPlayer. For example:

```
$ omxplayer file.mp3
```

If you need to play the sound file from within a Python program you can use the `os.system` command to run OMXPlayer, as follows:

```
import os
os.system('omxplayer file.mp3')
```

Discussion

In addition to the headphone jack, sound can also be played through the HDMI connector. The OMXPlayer software should automatically play through HDMI unless the audio jack has something plugged into it.

You can choose the channel the sound should be played through by using the `-o` option, which can have a value of either `local` (audio socket) or `hdmi`. For example:

```
$ omxplayer -o hdmi file.mp3
```

See Also

You can find out more about the OMXPlayer here: *http://elinux.org/Omxplayer*. In addition to sound files in most formats, the player can also play video files.

For an alternative method of playing sound files using the Pygame library, see *http://bit.ly/2mt4eDM*.

The Raspberry Pi Zero does not have an audio jack, but you can still use it to play audio; see *http://bit.ly/2mIjZH9* for details.

18.3 Incorporate an Electret Microphone Into a Project

Problem

You need a preamplifier for an electret microphone, either for further amplification to use speakers (power amplifier) or for detecting sound levels with an Arduino.

Solution

Use a single-supply rail-to-rail op-amp like the OPA365 and amplify the signal by a factor of between 30 and 100. A gain of 30 is about right if you will be speaking directly into the microphone and 100 if you want the microphone to be useful in picking up ambient sounds.

Figure 18-3 shows the schematic for the microphone preamp with a gain of 101. For a discussion of how to set the gain, see Recipe 17.5.

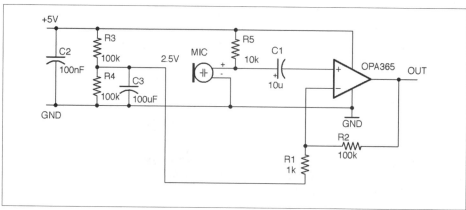

Figure 18-3. An Electret Microphone Preamplifier

C1 is used to "AC couple" the weak output from the microphone to the noninverting input of the op-amp. It allows only the AC part of the signal from the mic, rejecting the DC bias that would otherwise be present.

Discussion

Electret microphones operate like capacitors. Sound waves effectively move one plate of the capacitor closer or further away. The plates of the electret are charged during manufacture and retain that charge for the life of the microphone. These changes in capacitance are converted into a small voltage using an FET transistor built into the microphone. The microphone module is polarized and must be connected the right way around. You also need to supply a drain resistor for the FET (R5 in Figure 18-3).

The circuit of Figure 18-3 will produce an output centered around 2.5V that can be connected directly to an Arduino's analog input and allow you to do things like measure the maximum sound level.

Figure 18-4 shows the circuit used with an Arduino Uno.

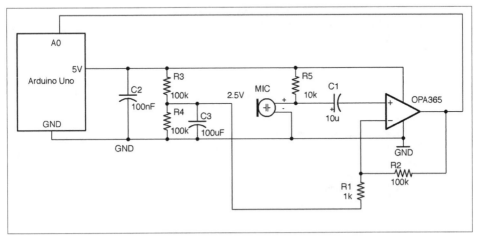

Figure 18-4. An Arduino Sound-Level Meter

This circuit can be built on a breadboard, but the OPA365 is only available as an SMD, so you will need to use a SOT-23 or SOIC breakout board such as the one from Schmartboard.com, as shown in Figure 18-5.

The Arduino sketch *ch_18_sound_meter* (see Recipe 10.2) samples the analog input A0 for 100ms and then reports on the maximum amplitude it sampled (Figure 18-6). You could modify this sketch to trigger some action if the sound level rises above a certain level.

Note that the maximum possible reading will be 512 (half the full analog value reading):

```
const int soundPin = A0;
const long samplePeriod = 100; // ms

long lastSampleTime = 0;
int maxAmplitude = 0;
int n = 0;

void setup() {
  Serial.begin(9600);
}
```

```
void loop() {
  long now = millis();
  if (now > lastSampleTime + samplePeriod) {
    processSoundLevel();
    n = 0;
    maxAmplitude = 0;
    lastSampleTime = now;
  }
  else {
    int amplitude = analogRead(soundPin) - 512;
    if (amplitude > maxAmplitude) {
      maxAmplitude = amplitude;
    }
    n++;
  }
}

void processSoundLevel() {
  // replace or add your own code to use maxAmplitude
  Serial.print("Of ");
  Serial.print(n);
  Serial.print(" samples, the maximum was ");
  Serial.println(maxAmplitude);
}
```

Figure 18-5. Using an SMT OPA365 on a Breadboard

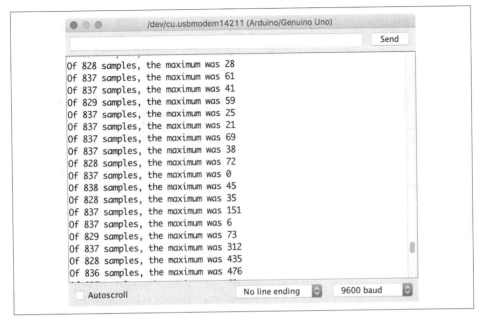

Of 828 samples, the maximum was 28
Of 837 samples, the maximum was 61
Of 837 samples, the maximum was 41
Of 829 samples, the maximum was 59
Of 837 samples, the maximum was 25
Of 837 samples, the maximum was 21
Of 837 samples, the maximum was 69
Of 837 samples, the maximum was 38
Of 828 samples, the maximum was 72
Of 837 samples, the maximum was 0
Of 838 samples, the maximum was 45
Of 828 samples, the maximum was 35
Of 837 samples, the maximum was 151
Of 837 samples, the maximum was 6
Of 829 samples, the maximum was 73
Of 837 samples, the maximum was 312
Of 828 samples, the maximum was 435
Of 836 samples, the maximum was 476

Figure 18-6. Serial Monitor Reporting Sound Level

See Also

Ready-made microphone preamplifier modules are available that use electret microphones, for example: *https://www.sparkfun.com/products/12758*.

Alternatively, you can use the type of microphone found in cellphones called MEMS
(microelectromechanical systems). These are essentially microphones on a chip.
SparkFun and others sell breakout boards for these extremely tiny microphones, such
as this one: *https://www.sparkfun.com/products/9868*.

For information on using analog inputs with an Arduino, see Recipe 10.12.

18.4 Make a 1W Power Amplifier

Problem

You need a low-cost power amplifier to power a small speaker at low volumes, say to
accompany an FM radio receiver as in Recipe 19.3.

Solution

Use a power-amp IC like the TDA7052, which will work with any power supply from
3V to 15V and can supply 1W of power into an 8Ω speaker. The schematic for a simple amplifier using this device is shown in Figure 18-7.

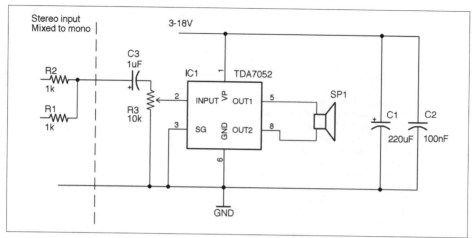

Figure 18-7. Using the TDA7052 1W Linear Power Amplifier

R1 and R2 are only provided to show how to mix the left and right signals of a stereo audio source to mono. If you have a mono source, just connect it directly to C3.

R3 acts as a volume control, and if the sound source has a volume control, you can replace this with two fixed resistors acting as a voltage divider.

C1 and C2 provide energy stores for the rapidly changing load of the IC as it drives the speaker. C1 should be placed as close to IC1 as possible.

Discussion

Figure 18-8 shows how you can build an amplifier like this on a solderless breadboard. In this case, the breadboard is stuck to a Monk Makes Protoboard for the convenience of being able to attach the speaker to screw terminals and use the audio jack and power jack.

What Counts as a "Power" Amplifier?

1W is not very powerful compared with the 20W or more you will generally find in a home stereo system. An amplifier is generally considered a "power" amp if it has enough current amplification to drive a low impedance load (i.e., 4 or 8Ω loudspeakers).

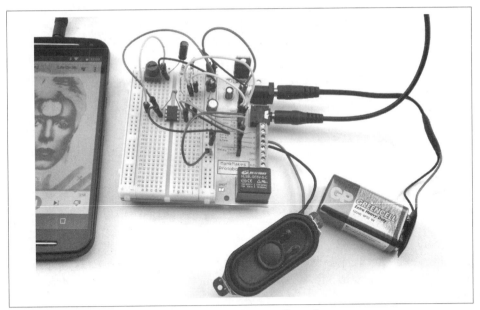

Figure 18-8. A 1W Power Amplifier Built on a Breadboard

See Also

The TDA7052 datasheet can be found here: *http://bit.ly/2nA0GgB*.

For high-power class-D design, see Recipe 18.5.

18.5 Make a 10W Power Amplifier

Problem

You want a low-cost efficient power amplifier that will run cool even at full power.

Solution

Use a class-D digital amplifier IC such as the TPA3122D2. The schematic for this is shown in Figure 18-9.

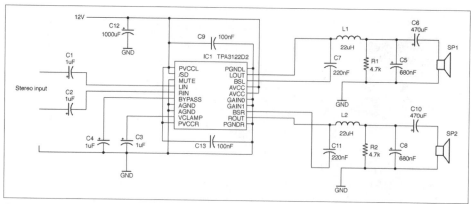

Figure 18-9. A Class-D Power Amplifier Using a TPA3122D2

This design is a modified version of the one found on page 14 of this chip's datasheet. It will operate on a supply voltage of 10 to 30V, but at a convenient 12V will supply about 7.5W per channel into 4Ω speakers. For more power you need to increase the supply voltage.

The TPA3122D2 is a 20-pin DIL (dual in-line) IC that makes the design very suitable for prototyping on a breadboard (Figure 18-10).

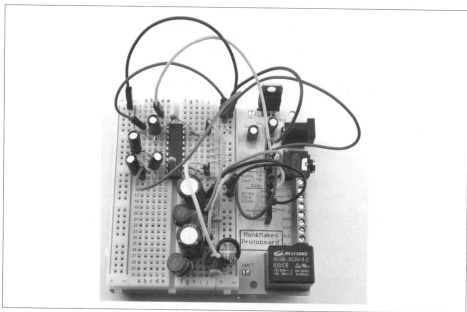

Figure 18-10. A Class-D Power Amplifier Using a TPA3122D2 on a Breadboard

One interesting feature of the TPA3122D2 is that the gain is set by two digital inputs that set the gain to one of four values. Table 18-1 shows how these pins are used. In the schematic of Figure 18-9 the lowest gain of 20dB is selected.

Table 18-1. Setting the Gain of a TPA3122D2

GAIN1	GAIN0	Amplifier gain (dB)
LOW	LOW	20
LOW	HIGH	26
HIGH	LOW	32
HIGH	HIGH	36

Discussion

Unlike traditional amplifier designs like that of Recipe 18.4 class-D amplifiers are extremely efficient, transferring sometimes over 90% of the power supplied to them to the output load (loudspeaker).

Figure 18-11 shows how a class-D amplifier works.

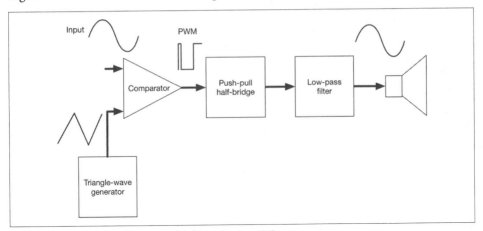

Figure 18-11. Block Diagram of a Class-D Amplifier

The input signal is converted into high-frequency PWM (250kHz in the case of the TPA3122D2) by a combination of a triangle-wave generator and comparator. This is actually the way Recipe 16.9 works. The length of the pulse will be determined by the time it takes for the triangular signal to exceed the value of the analog input. Since the analog input is a much lower frequency than the triangle wave, it can be considered as having a constant instantaneous value, while the triangle wave is "sampling" it. It's a bit like driving fast past someone who is walking—the pedestrian is effectively stationary from the car's perspective.

This PWM signal can then be amplified by transistors in a half-bridge switching arrangement such as the one in Recipe 11.8. This is then low-pass filtered and fed to the loudspeaker.

Digital amplifiers are perfectly acceptable for many applications, but they are unlikely to be accepted by HiFi enthusiasts without considerable improvements in design. They generally have higher levels of distortion than analog amplifiers. In Figure 18-12 you can clearly see the distortion in the sinewave being amplified.

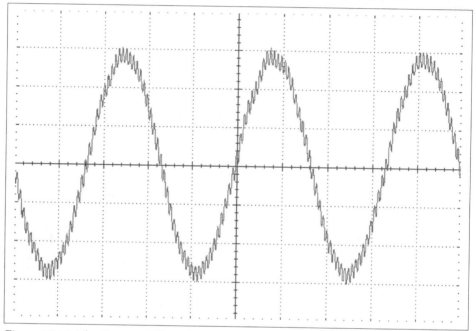

Figure 18-12. The Output of This Class-D Amplifier with a 6kHz Sinewave Input

See Also

For the TPA3122D2 datasheet see *http://www.ti.com/lit/ds/symlink/tpa3122d2.pdf*.

## CHAPTER 19

Radio Frequency

19.0 Introduction

It used to be that any electronics book worth its salt would include discrete circuits for AM and FM broadcast receivers and maybe even discuss how an analog TV works. In these digital days, topics like this are only for historical interest.

For this reason, this chapter mostly concentrates on the use of digital communications using radio frequencies. The discussion of the nature of radio is restricted to this introduction. The recipes themselves are fundamentally practical.

We take radio frequency for granted now, but when it first became popular, it must have seemed like magic. After all, it allowed you to talk to someone miles away without any wires connecting them. How could this be possible?

Transmitters and the Law

Recipe 19.1, Recipe 19.2, and Recipe 19.4 are radio transmitters. Although most countries allow the use of low-power, short-range FM transmitters and allocate certain bands for use with packet radio, you should check your local regulations and make sure you are not breaking the law by using any of these designs.

Amplitude Modulation (AM)

It all starts with AM transmissions, which use a carrier frequency modulated by the audio signal that is being transmitted. Figure 19-1 shows how this works.

In Figure 19-1 the carrier frequency is only about 4.5 times the audio signal. In a real AM broadcast, the audio signal will generally be a maximum of 16kHz against a carrier frequency of 500kHz, making at least ten times as many cycles of carrier in each cycle of the audio frequency shown in Figure 19-1. But it's easier to see what is going on with fewer cycles.

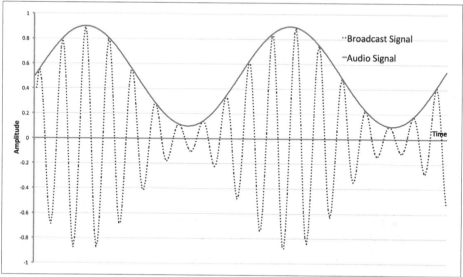

Figure 19-1. AM Transmission

Each transmitting station has its own frequency, and for AM public broadcasts in the medium wave (MW) band, this range is around 520kHz to 1600kHz. The amplitude of the carrier is modulated by the audio signal and the resulting signal broadcast through an antenna.

One of the reasons the quality of AM is so bad is that there are many other factors that can affect the amplitude of the carrier, including atmospheric conditions or changes in position of the receiver if you are listening in a moving vehicle.

When it comes to the receiver, you need to be able to separate the radio-frequency carrier (RF) from the audio signal.

The first step in this is to "tune in" to the transmission frequency you are interested in. This means using a narrow band-pass filter (Recipe 17.9) so that all the other stations get tuned out. This is traditionally accomplished by using a tuned circuit in the form of a fixed inductor (also acting as the antenna) and a variable capacitor that is varied to tune into different frequencies.

Since your ears cannot hear frequencies above 20kHz (lower as you get older) a radio carrier frequency of 500kHz or more will be inaudible. With your ears acting as a low-pass filter, you might expect that with suitable amplification of the tuned radio signal, you would be able to hear the audio signal modulating it. This is almost correct, but because the "average" of each cycle (whatever the amplitude) is zero (each positive wave followed by a negative one), you won't hear anything. However, if you introduce a diode to remove one half of the signal (Figure 19-2), then suddenly the original audio signal can be separated and you will be able to hear the audio in the envelope of the AM signal and your ears will now happily be a low-pass filter for you.

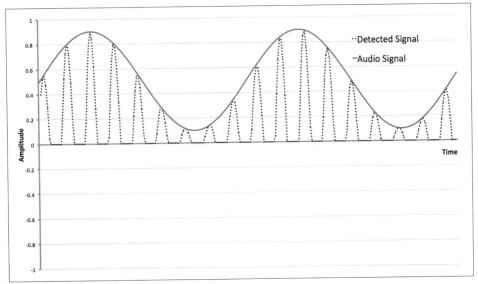

Figure 19-2. Detecting the Audio Signal from AM

The big empty space on the bottom of Figure 19-2 serves to show where the diode has removed the negative part of the tuned signal.

The earliest AM receivers (crystal sets) were literally a coil, antenna, variable capacitor, diode (germanium for its lower forward voltage), and sensitive headphones as shown in Figure 19-3.

Improvements have been made to the design of AM receivers, in particular amplification of the RF signal before detection, followed by further amplification to drive a loudspeaker, but the basic principles of an AM receiver are as they were a century ago.

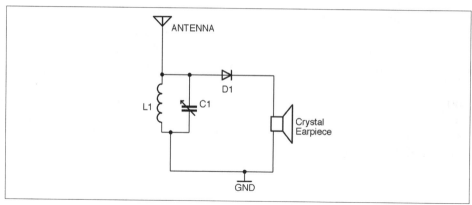

Figure 19-3. A Basic AM Receiver

Frequency Modulation (FM)

Frequency modulation (FM) improves on AM by varying the frequency of the carrier rather than its amplitude. This improves the quality of the sound being broadcast, as the frequency is immune from changes due to atmospheric conditions or movement of the receiver. Figure 19-4 shows how an audio signal would be used to modulate a carrier.

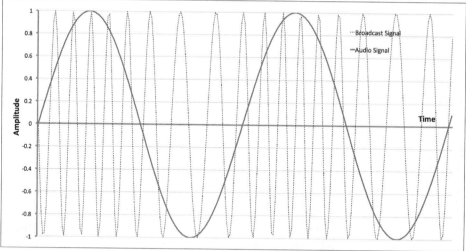

Figure 19-4. Frequency Modulation

To create an FM transmitter, a VCO (Recipe 16.10) is used. The audio signal is fed into the frequency control input of the VCO to vary the frequency. The first recipe in this chapter (Recipe 19.1) is a simple, low-power FM transmitter.

Extracting the audio signal from an FM broadcast can be accomplished by using a number of techniques, but perhaps the most common is to use a VCO in an arrangement called a phase-locked-loop (PLL). Figure 19-5 shows a PLL configured as an FM demodulator.

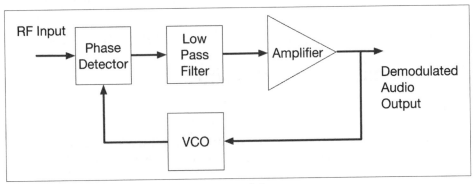

Figure 19-5. A Phase-Locked Loop FM Demodulator

The RF signal is first tuned to a particular frequency as with an AM receiver, but then to demodulate the original audio, the RF signal is fed into a phase comparator. The phase comparator produces an output that is zero if its two input signals are completely in phase. If the signals are out of phase with each other then the comparator produces an output proportional to the difference in phase. This signal is then low-pass filtered, amplified, and used to control a VCO whose output supplies the second input to the phase comparator. This feedback results in the VCO producing a signal of the same frequency as the RF input to the phase comparator. As the output of the VCO tracks the RF frequencies the small shifts in phase resulting from the modulation will result in the audio signal being available in the filtered and amplified output of the phase detector. The amplified output contains the demodulated audio signal almost as a side effect of the PLL tracking the broadcast signal.

PLL chips are available that include the VCO, phase comparator, and other parts all in one convenient package. However, if you are making an FM receiver, then you would just use an FM receiver IC that includes the RF amplification, PLL, and pretty much everything needed for the receiver aside from a few resistors, capacitors, and inductors.

Digital Radio

Radio is now very much a digital affair. FM radio receivers and transmitters are now often implemented as software-defined radios (SDRs). Processors are now so fast that with a very small amount of hardware they can produce and decode analog broadcast signals in software. For example, the filtering, phase detecting, etc. of a PLL can all be done as software algorithms rather than directly in hardware. In Recipe 19.2 you will find a recipe for a SDR FM transmitter using a Raspberry Pi.

In addition to using digital electronics to decode analog signals, the use of digital signals carried on RF is everywhere from your cellphone to the remote you use to unlock your car.

This chapter has several recipes for passing digital data wirelessly between devices.

19.1 Make an FM Radio Transmitter

Problem

You want to make a short-range FM transmitter that will transmit the sound signal on a domestic FM frequency where it can be received on a household receiver.

Solution

Use a high-frequency VCO chip (MAX2606) set to operate in a frequency on the FM band and use the audio signal to make small adjustments to the frequency.

The schematic for this is shown in Figure 19-6.

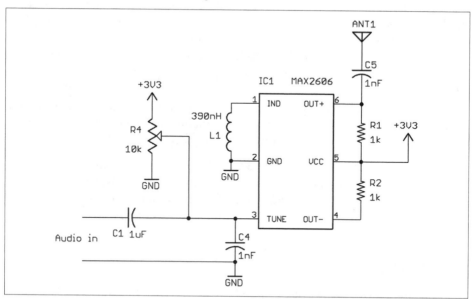

Figure 19-6. An FM Transmitter Using a VCO IC

The variable resistor R4 is used to tune the transmitter to a particular frequency. The antenna can be a telescopic FM antenna or simply a few feet of wire.

Discussion

The inductor L1 sets the center frequency of the VCO. The datasheet for this device indicates that an inductor of 390nH will give a center frequency of around 102MHz.

The control signal applied to the TUNE pin of the VCO comprises a fixed DC offset supplied by R4 (to allow tuning to a particular transmission frequency) and an audio signal that is sufficient to produce the desired frequency modulation.

The output OUT+ of the MAX2606 drives the antenna. The datasheet also suggests that complementary open-collector output OUT– is also supplied with a 1kΩ pull-up resistor.

See Also

This recipe is based on the tutorial by Afroman (*http://bit.ly/2n302r1*) who also has an open source hardware circuit board that you can make yourself or buy from OSH-Park (*http://bit.ly/2mtaNWV*).

The datasheet for the MAX2606 can be found at *http://bit.ly/2ltwp5O*.

19.2 Create a Software FM Transmitter Using Raspberry Pi

Problem

You want to know how you can use your Raspberry Pi to act as an FM transmitter.

Solution

Use the PiFM software and attach an antenna to GPIO pin 4 of your Raspberry Pi. This can be just a female-to-male jumper wire as shown in Figure 19-7.

First, you'll need to download and install PiFM using the following commands:

```
$ mkdir pifm
$ cd pifm
$ wget http://omattos.com/pifm.tar.gz
$ tar -xvf pifm.tar.gz
```

If you have a Raspberry Pi 2 or 3, you have to download and compile a modified version of the software to work on the newer hardware, so run the following commands:

```
$ git clone https://github.com/oatmeal3000/pi2fm.git
$ mv pi2fm pi2fmdir
$ mv pi2fmdir/pi2fm.c .
$ gcc -lm -std=c99 -g pi2fm.c -o pi2fm
```

Figure 19-7. A Raspberry PiFM Transmitter

You will now have two executable programs, one called *pifm* for the Raspberry Pi 1 models and *pi2fm* for the Raspberry Pi 2 and 3. If you have an original Pi 1 you will need to modify *pi2fm* to be just *pifm* in the following commands, which will play the file *sound.wav* supplied with the *pifm* software:

```
pi@raspberrypi:~/pifm $ sudo ./pi2fm sound.wav 94.0
starting...
 -> carrier freq: 94.0 MHz
 -> band width: 8.0
now broadcasting: sound.wav ...
```

Discussion

Attaching a longer wire to pin 4 will considerably increase the range of the transmitter.

See Also

The original web page for the PiFM project can be found here: *http://bit.ly/18AcT5u*.

Details on the Raspberry Pi 2 version are here: *https://github.com/oatmeal3000/pi2fm*.

19.3 Build an Arduino-Powered FM Receiver

Problem

You want to make an FM radio receiver controlled by an Arduino.

Solution

Use a TEA5767 FM radio-receiver module controlled by an Arduino and either connect headphones or use a power amplifier to drive speakers.

Figure 19-8 shows how the module is connected to an Arduino.

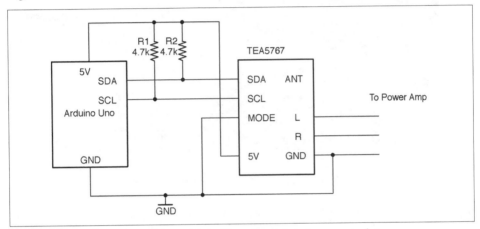

Figure 19-8. Connecting a TEA5767 FM Receiver Module to an Arduino

To make the module as easy to program as possible, download the Arduino TEA5767 library by clicking on Clone or download on GitHub (*https://github.com/simonmonk/arduino_TEA5767*). Select the option to download as a ZIP file and then from the Arduino IDE select the menu option Sketch→Include Library→Add ZIP Library and navigate to the ZIP file you just downloaded.

Discussion

To build a circuit using this module on a breadboard (Figure 19-9), you will need a breakout circuit board that allows you to connect the fine pitch of the module's connectors to the breadboard. You can make this yourself using stripboard (*http://bit.ly/2mbOYKh*) or buy a breakout board from OSHPark (*http://bit.ly/2ltxNFx*) or Monk Makes (*http://bit.ly/2mUg8E2*).

Figure 19-9. An Arduino-Controlled FM Radio on a Breadboard

The sketch *ch_19_fm_radio* uses the Serial Monitor so that you can enter the frequency you want to tune in to. When using the Serial Monitor with this sketch, make sure the "Line ending" drop-down list is set to "No line ending" before sending a frequency to the Arduino.

The sketch is available with the book downloads (see Recipe 10.2) and is listed here:

```
#include <Wire.h>
#include <TEA5767Radio.h>

TEA5767Radio radio = TEA5767Radio();

void setup() {
  Serial.begin(9600);
  Serial.println("Enter Frequency:");
  Wire.begin();
}

void loop() {
  if (Serial.available()) {
    float f = Serial.parseFloat();
    radio.setFrequency(f);
    Serial.println(f);
```

```
    }
}
```

The `setFrequency` function accepts a decimal value in MHz (e.g., 93.0).

See Also

To make a suitable power amplifier for this project, see Recipe 18.4 or Recipe 18.5.

For an FM transmitter to accompany this receiver, see Recipe 19.1 or Recipe 19.2.

19.4 Send Digital Data Over a Radio

Problem

You want to send data hundreds of yards over a radio connection.

Solution

Use a CC1101 RF transceiver (transmitter/receiver). These boards are readily available on eBay at very low cost. Figure 19-10 shows how you would wire one up to the SPI interface of an Arduino.

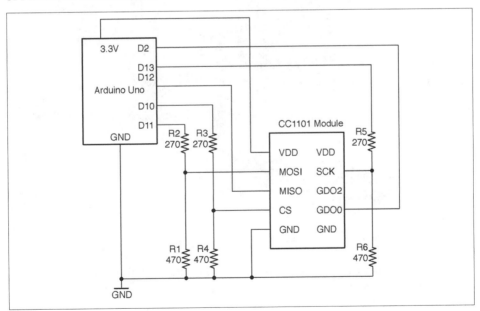

Figure 19-10. Wiring a CC1101 Module to an Arduino

The module is a 3.3V device and the datasheet states that none of the pins should have a voltage of more than 3.9V applied to them, so a level converter should be used on all pins used as inputs to the CC1101. This takes the form of six resistors used in pairs as voltage dividers (see Recipe 2.6).

Discussion

Clearly to test out this recipe, you will need two Arduinos and two CC1101 modules each wired up as shown in Figure 19-11.

Figure 19-11. An Arduino Uno Wired to a CC1101 Module

Testing the modules can get very confusing because the Arduino IDE can only have one port selected at a time. It's much easier to find a second computer to use both for programming the transmitter and for running the Serial Monitor.

The library used for these examples should be downloaded as a ZIP file from *https://github.com/simonmonk/CC1101_arduino* by clicking on the Clone or download button and selecting Download ZIP. Save the ZIP file somewhere and then from the

Sketch menu choose Sketch→Include Library→Add ZIP Library and select the ZIP you just downloaded.

The two example programs (one for transmit and one for receive) are the examples that accompany the library, but these can also be downloaded with the downloads for the book (see Recipe 10.2).

Here is the transmitter code (*ch_19_cc1101_tx*):

```
#include <ELECHOUSE_CC1101.h>

const int n = 61;
byte buffer[n] = "";

void setup() {
  Serial.begin(9600);
  Serial.println("Set line ending to New Line in Serial Monitor.");
  Serial.println("Enter Message");
  ELECHOUSE_cc1101.Init(F_433); // set frequency - F_433, F_868, F_965 MHz
}

void loop() {
  if (Serial.available()) {
    int len = Serial.readBytesUntil('\n', buffer, n);
    buffer[len] = '\0';
    Serial.println((char *)buffer);
    ELECHOUSE_cc1101.SendData(buffer, len);
  }
}
```

The maximum packet size is 64 bytes and `buffer` is used to contain the data to be sent, which can by anything you can pack into a byte array. In this case, short text messages.

In the `setup` function, the communication frequency is set. The Serial Monitor first reminds you to turn line endings on in the Serial Monitor. When you send a message the contents of the message are written into `buffer` and `SendData` then used to transmit the message you typed.

The corresponding receiver code can be found in *ch_19_cc1101_tx* and is listed here:

```
#include <ELECHOUSE_CC1101.h>

const int n = 61;

void setup()
{
  Serial.begin(9600);
  Serial.println("Rx");
  ELECHOUSE_cc1101.Init(F_433);  // set frequency - F_433, F_868, F_965 MHz
  ELECHOUSE_cc1101.SetReceive();
}
```

```
byte buffer[61] = {0};

void loop()
{
  if (ELECHOUSE_cc1101.CheckReceiveFlag())
  {
    int len = ELECHOUSE_cc1101.ReceiveData(buffer);
    buffer[len] = '\0';
    Serial.println((char *) buffer);
    ELECHOUSE_cc1101.SetReceive();
  }
}
```

This sketch uses `CheckReceiveFlag` to monitor Arduino pin 2 for the CC1101 to indicate that a new message has arrived. When a message does arrive, it is read into `buffer` and then displayed in the Serial Monitor. Figures 19-12 and 19-13 show the Serial Monitors of the transmitter and receiver, respectively.

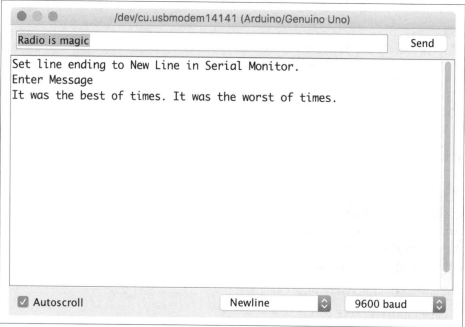

Figure 19-12. Transmitting a Message Using a CC1101 Module

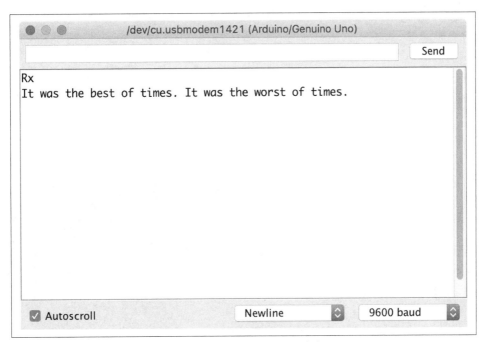

Figure 19-13. Receiving a Message Using a CC1101 Module

See Also

You can find the datasheet for the CC1101 IC here: *http://bit.ly/2ltuxK9*.

The library used here is modified from that of elechouse.com that also sells CC1101 modules: *http://bit.ly/2n3lnQK*.

## CHAPTER 20

Construction

20.0 Introduction

This chapter deals with practical techniques for making solderless prototypes and then building more durable versions of them.

20.1 Create Temporary Circuits

Problem

You want to construct a circuit quickly and easily without having to do any soldering.

Solution

Using a solderless breadboard (Figure 20-1) is a great way to temporarily build an electronic design because you can easily swap out different components without having to desolder (Recipe 20.6). Once the design is right it can be transferred either to a prototyping board (Recipe 20.2) or a circuit board of your own design (Recipe 20.3).

Breadboards are available in many shapes and sizes. I recommend having a good supply of 400-point breadboards (often called half-breadboards) as these are a great size to accommodate a couple of chips and a whole load of components. For those situations where you need more prototyping area breadboards can be clipped together.

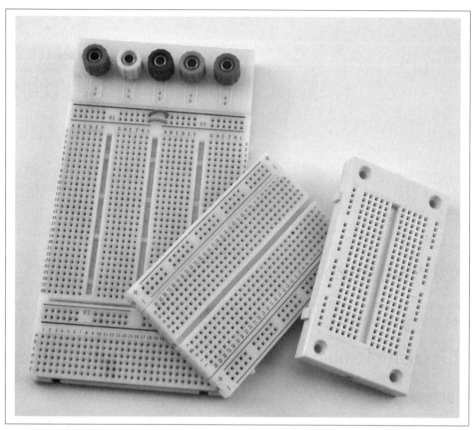

Figure 20-1. A Selection of Breadboards

Figure 20-2 shows how a 400-point breadboard is constructed of rows of five holes, all connected together by a clip. Note that there is a break in connectivity between "abcde" and "fghij."

The "supply rails" in pairs down the sides of the board are all connected together and can be used for any purpose, but generally the ones marked in blue or with a "–" next to them are used for ground and the red ones marked with a "+" are used for the positive supply.

The rows and columns of the breadboard are labeled with numbers and letters, making it easier to transfer the designs to a more permanent design using Permaproto or Monk Makes Protoboard. These are general-purpose circuit boards with the same layout as breadboards.

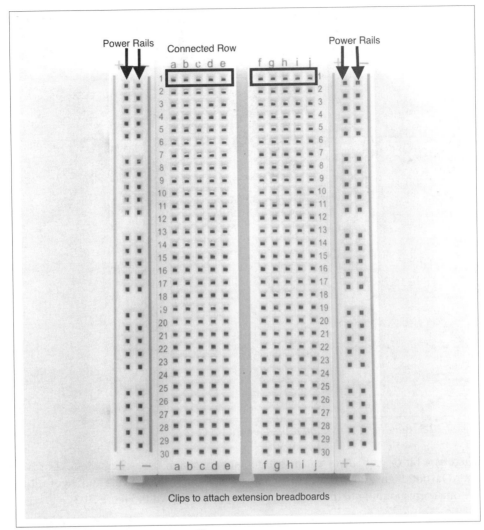

Figure 20-2. The "Half-Breadboard" Layout

Figure 20-3 shows the two-transistor oscillator from Recipe 16.2 built on a breadboard.

Through-hole component leads are pushed into holes on the breadboard and connected to other component leads either by being in the same breadboard row, or by using male-to-male jumper wires to link one part of the board to another.

Figure 20-3. Building a Circuit on a Breadboard

There is a bit of craft involved in the mental conversion of a schematic to a bread-board layout. Figure 20-4 shows the schematic layout for Recipe 16.2 and Figure 20-5 the one corresponding to the breadboard layout (also shown in Figure 20-3).

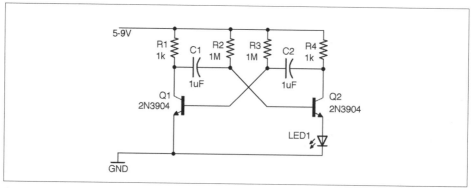

Figure 20-4. Schematic for a Transistor Oscillator

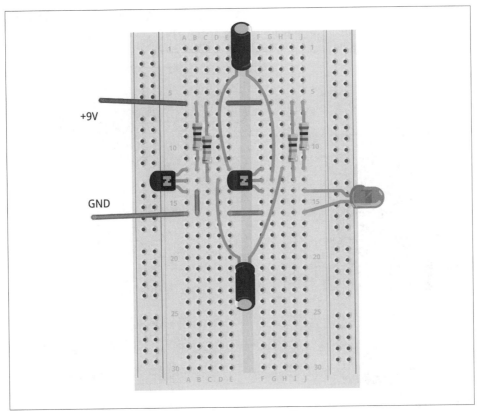

Figure 20-5. Breadboard Layout for a Transistor Oscillator

When transferring a schematic to a breadboard, I start with the key components and the ones whose legs won't bend very far. Usually this means starting with any ICs, but in this case, it means the transistors, which I have placed on opposite halves of the breadboard, mirroring the schematic.

Then I add the other components and jumper wires. There is no hard and fast method for this layout, and of course, being solderless, it's very easy to change things around if you run out of room.

Discussion

Figure 20-5 was drawn using an open source software tool called Fritzing. This tool produces great-looking diagrams and also allows you to view the same circuit as a schematic, breadboard, or circuit board design, allowing you to use the same software right through from prototyping to circuit board.

When it comes to connecting a breadboard to an Arduino or Raspberry Pi, you can use male-to-male or male-to-female jumper wires, respectively. Figure 20-6 shows Recipe 12.12 on a breadboard connected to a Raspberry Pi.

Figure 20-6. Connecting a Breadboard to a Raspberry Pi

Figure 20-7 shows an Arduino Micro on a solderless breadboard. This is one of several models of Arduino including the "mini" and "nano" models that are "breadboard friendly."

The breadboard shown in Figure 20-3 is likely to have proper electronic engineers holding their heads in their hands and gritting their teeth. The problem with it is that it is all too easy for a component lead to touch another lead that it shouldn't or simply fall out. This type of arrangement requires care to make sure nothing is touching that shouldn't be but is fine while you are getting something working. For a halfway house before you move to a soldered prototype, you can take a neater approach and use solid-core wire trimmed to just the right length to lie flat on the breadboard instead of using jumper wires. You can also trim all the component legs so that they are just the right lengths, but this can make them difficult to reuse in your next design.

Figure 20-7. An Arduino Micro on a Breadboard

See Also

For more information on Fritzing see *http://fritzing.org* or my book *Fritzing for Inventors*, TAB DIY, 2015.

When you are ready to transfer your design to a soldered prototype follow Recipe 20.2.

20.2 Create Permanent Circuits

Problem

You want to transfer your solderless breadboard design to a soldered design, either as a more durable prototype or as the final build for a one-off project.

Solution

Use a prototyping board or protoboard such as the Adafruit Permaproto or the Monk Makes Protoboard and simply transfer the components from their positions on the breadboard to the same locations on the protoboard.

Figure 20-8 shows the starting point for this process of transferring the design from a breadboard to a Monk Makes Protoboard. The details of the design are not important, but for those interested, it is an energy monitor using a particle photon Arduino-like board, a current sensor, and an AC transformer. Note that the breadboard itself is attached to a Monk Makes Protoboard to allow use of the screw terminal and audio-type socket to connect the current sensor and AC transformer to the prototype.

Figure 20-8. The Prototype on a Breadboard

The components can then be moved from the breadboard one at a time, making note of the lead's row and column coordinates and soldering them onto the Monk Makes Protoboard as shown in Figure 20-9. Finally, Figure 20-10 shows the completed board and the bare breadboard ready for your next project.

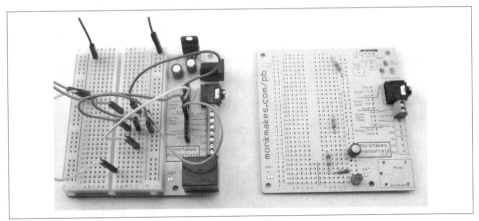

Figure 20-9. Transferring the Components to the Protoboard

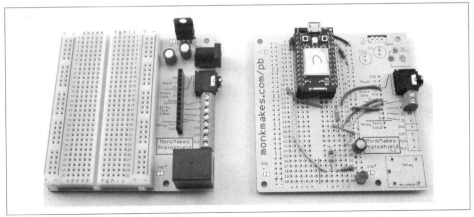

Figure 20-10. The Completed Protoboard

Discussion

Other useful prototyping boards are available including Permaproto boards from Adafruit that are designed to be attached by a ribbon connector to a Raspberry Pi.

When it comes to Arduino, the Arduino Protoshield is a great way to build something and then plug it as a whole board into the top of an Arduino Uno. Figure 20-11 shows a Protoshield with an LED cube project built onto it. The Protoshield is then attached to an Arduino.

Figure 20-11. An Arduino Protoshield LED Cube

Another popular method for making soldered prototypes is to use stripboard. Stripboard (Figure 20-12 and Figure 20-13) is arranged with copper tracks in rows and can be cut to whatever size you need for your project. Breaks in the track are made using a tool called a "spot cutter" or by using a drill bit rotated by hand just enough to cut the copper.

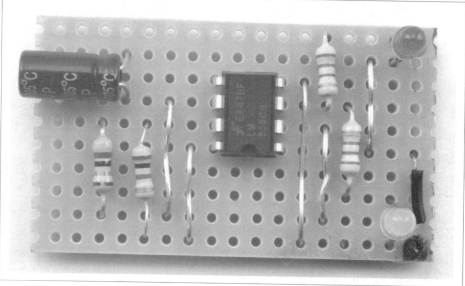

Figure 20-12. Stripboard (top side)

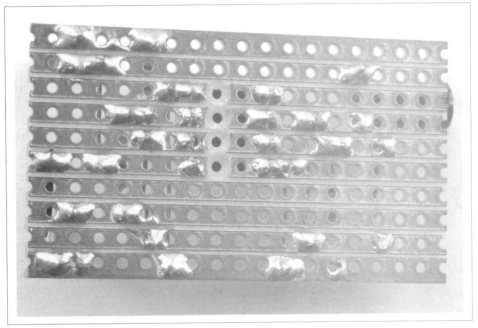

Figure 20-13. Stripboard (under-side)

Perfboard (i.e., perforated board) is like stripboard, but without the copper strips on the back. All connections are made by bending the component leads or lengths of solid-core wire.

Protoboard, which is a variation on perfboard with a separate pad behind each hole, is also available.

See Also

For Permaproto see: *https://www.adafruit.com/product/571.*

For Monk Makes Protoboard see: *https://www.monkmakes.com/pb/*

The LED cube project is taken from my book *The TAB Book of Arduino Projects*, TAB DIY, 2015.

20.3 Design Your Own Circuit Board

Problem

You want to design a circuit board for your project.

Solution

While the solution to this recipe cannot fit into one recipe, you can start by using circuit board design software such as CadSoft EAGLE to draw the schematic for your boards and then layout a circuit board. You can then send away the design files (known as Gerbers) to a circuit board fabrication provider.

Designing a simple circuit board is well within the reach of the amateur. If you restrict yourself to two copper layers, then you can use one of many CAD (computer-aided design) packages for circuit board design.

Perhaps the most popular CAD system is EAGLE CAD (Figure 20-14), which is not open source, but there is a free version for noncommercial projects. One advantage of using EAGLE CAD is that it has been adopted by the open source hardware movement as their CAD system of choice and there are many OSH designs for which the EAGLE design files are available for download, which you can adapt to your own needs. EAGLE is not an intuitive piece of software. If you have used any other circuit board design software, you may be able to find your way around unaided, but typically you will need to follow a few tutorials step-by-step before it really starts to make sense.

Nearly all the schematic diagrams in this book were drawn with EAGLE and once you are familiar with all its quirks, you will probably grow to enjoy using it.

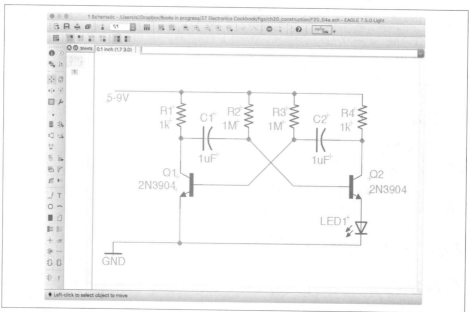

Figure 20-14. The EAGLE CAD Schematic Design Editor

For open source enthusiasts KiCad is a fully featured open source equivalent to EAGLE, and it can import some EAGLE file formats. Like EAGLE, don't expect to have a finished board design ten minutes after downloading the software.

DesignSpark is a free-ish CAD system that is gaining in popularity and is relatively easy to use, but it does inflict adverts on users.

If you have a really simple circuit board design to make that uses fairly common components, then Fritzing will take you to circuit board design via schematic and breadboard and is easy to use. However, if you run into parts that are not included with Fritzing (and you probably will) you may well have to make your own parts, which is a little tricky and requires you to have some skill editing multilayer SVG files.

All of the CAD options described will give you the option to export Gerber design files. You can then use an online circuit board service that will let you upload a ZIP file containing the design, and in a week or two, send you back a small batch of boards. The minimum is usually 10 for as little as a dollar a board for small boards.

The circuit board services are changing all the time. I have used PCBWay, ITEAD Studio, and Seeed Studio for circuit board manufacture without any problems.

Discussion

It used to be worth making your own circuit boards. In fact, I still have all the equipment for this tucked away in a box that I don't expect ever to open. Photo etching your own circuit boards does not need to be expensive; you can make your own UV exposure box with UV LEDs easily enough, but you will also need noxious chemicals that do not keep once mixed and are difficult to dispose of responsibly. The end results are also generally inferior to the circuit boards a professional service will produce for you. So my advice is to plan ahead a little and have your circuit boards produced professionally.

See Also

For a guide to EAGLE CAD see: *Make Your Own PCBs with EAGLE: From Schematic Designs to Finished Boards*, TAB DIY, 2014.

The KiKad website is here: *http://kicad-pcb.org/*.

You can find out about DesignSpark here: *https://www.rs-online.com/designspark/pcb-software*.

I also have also written a guide to using Fritzing called *Fritzing for Inventors: Take Your Electronics Project from Prototype to Product*, TAB DIY, 2015.

20.4 Explore Through-Hole Soldering

Problem

You want to learn the best technique for soldering through-hole components onto a circuit board.

Solution

Make the junction of the component lead and the solder pad hot for a second or two before applying the solder, as shown in Figure 20-15.

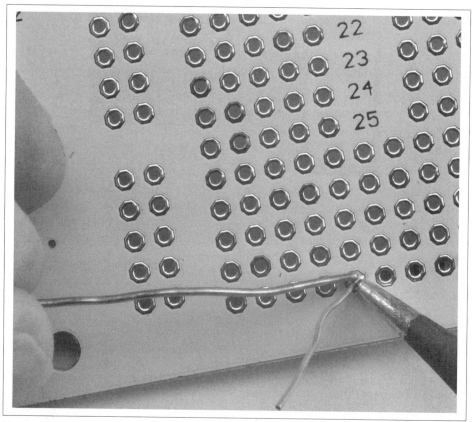

Figure 20-15. Soldering a Circuit Board Joint

In more detail, to make a good solder joint, you should:

1. Make sure your iron is up to temperature for the solder you are using. For leaded solder, that should be around 280°C (536°F) and for unleaded solder 310°C (590°F). If things aren't working well try altering the temperature.

2. Clean the tip of your iron on a damp sponge or brass wool. It should look shiny when you are done.
3. Touch the iron to where the component lead meets the solder pad and leave it there for just a second or so to heat.
4. Feed in some solder to where the circuit board and component lead meet until the solder has flown around the lead covering all the pad, with a little peak of solder pulled up toward the top of the lead.
5. Cut off the excess lead.

Safe Soldering

Soldering irons get hot—very hot—and can inflict nasty burns, so be careful. Safety glasses are a good idea as occasionally specks of solder can fly off if flicked by a lead, or even from boiling flux, and if your eyes are close to the thing you are soldering (which is quite likely) you could damage your eyes.

It is also a good idea to use a fume extractor (these do not need to be expensive; see Appendix A) to take away the noxious fumes released as the flux in the solder burns. This is not really something you want in your lungs.

Discussion

Due to various international regulations that have been put in place, lead (the element Pb) is rapidly being phased out of use in circuit board manufacture. Most solder is now unleaded. Unfortunately, unleaded solder has a higher melting point and is generally a bit more difficult to use than the leaded variety. You can still buy leaded solder and many people keep a reel of "the good stuff."

If you are using small components, then use a thin multicore rosin solder. I use 0.7mm.

See Also

You will find some good video tutorials for soldering on YouTube, and SparkFun has a great video on how to solder here: *http://bit.ly/2mDxVyD*.

I also recommend the book *Learn to Solder* by Brian Jepson, Maker Media, 2012.

20.5 Explore Surface-Mount Soldering

Problem

You want to solder a SMD to a circuit board.

Solution

If you are making a one-off, or just a few prototypes, consider soldering by hand. Unless you are very skilled, the results will not be great, but you should be able to make a functional prototype this way.

Soldering by hand is a great deal easier if you stick to SMD resistors and capacitors that are 0603 or larger and ICs with a pin spacing of 50 mils (1.27mm) or wider as found in SOIC packages.

Most components are available in a variety of different package sizes.

When soldering SMDs you will need to hold the part down (I use tweezers) while you solder each pin to a pad. As with through-hole soldering, heat the pin and pad for a second or so before touching the solder onto the pad and pin, flowing the solder into the joint.

When soldering the first pin, it's best to put a tiny blob of solder on one pad (Figure 20-16) and then solder the first pin by just heating and pressing the pin onto the pad (Figure 20-17). You should find that once one pin is soldered, you don't need to hold the device in place (Figure 20-18). When you have soldered all the other pins, it's a good idea to circle back to the first pin and just flow a bit of solder into it to make sure it's well soldered.

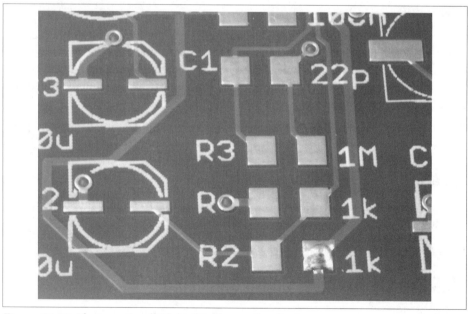

Figure 20-16. Placing a Small Blob of Solder on One Pad

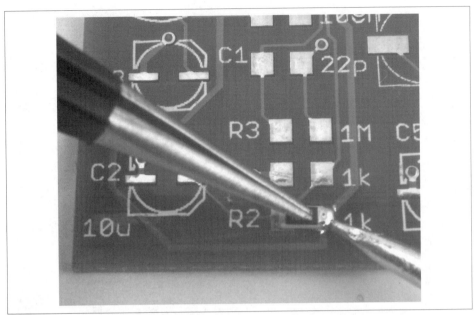

Figure 20-17. Holding the Component in Place While Soldering the First Pin

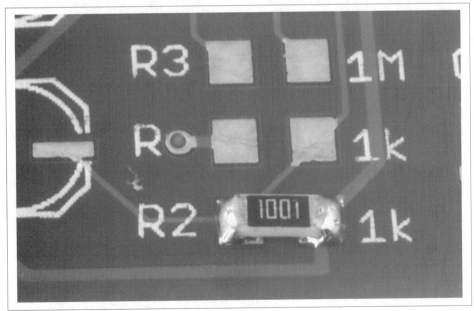

Figure 20-18. A Hand-soldered SMD Resistor

An alternative to using a soldering iron is to use solder paste and a hot-air gun. Here the procedure is to first place a little solder paste onto the component pads

(Figure 20-19). Place the component onto the pads and hold it down with the tip of the tweezers while you heat it up with the hot-air gun (Figure 20-20).

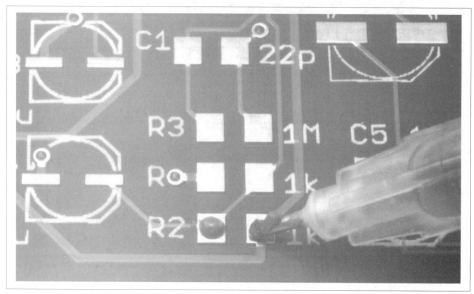

Figure 20-19. Placing a Small Blob of Solder Paste on the Pads

Figure 20-20. Holding the Component Down While Heating with the Hot-air Gun

If you have a reflow oven, like the homemade one shown in Figure 20-21, then you start as you would when using a hot-air gun, placing solder paste onto each pad. This task can be made easier by using a stencil, which most circuit board manufacturers will provide during circuit board manufacture for an extra cost.

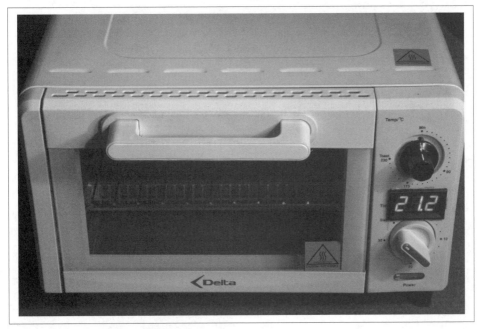

Figure 20-21. A Homemade Reflow Oven

Having applied the solder paste, place the components (they will stick to the solder paste). Then "cook" the board to melt the solder paste. A sophisticated reflow oven will have profiles you can set for different types of solder paste. It is surprising what you can get away with using a homemade oven. Mine is made from a converted toaster oven with a temperature probe fitted. A bit of trial and error will eventually get things right, but, purely as an example, my procedure for using lead solder paste (which has a lower melting temperature) is:

1. Place the board in the oven.
2. Put the oven on maximum until the temperature reaches 80°C, then turn it off and let it "soak" for 2 minutes during which the reported temperature will continue to increase, peaking at about 130°C.
3. Put the oven on maximum again until the solder paste has melted (at about 140°C) and then the oven should be turned off.
4. After a further 30 seconds, open the oven door to let the circuit board cool quickly.

Your procedure will no doubt be different, but when you've found the magic formula for your setup, it should work every time.

Homemade SMD Ovens Can Cause Fires

Modified toaster ovens are potentially very dangerous (they have the nickname "fire starters" for good reason). Both when making modifications to the oven and when using it you must know exactly what you are doing and the risks involved. If you are not absolutely sure about what you're doing, do not be tempted to make your own.

A homemade oven should not be left unattended for a moment while in use.

Discussion

Many circuit board fabricators also provide a circuit board assembly service. For a small batch of 5 or 10 prototypes, this can still be pretty cost effective as an alternative to cooking your own boards. It's certainly a lot safer than making your own reflow oven.

See Also

For soldering through-hole components, see Recipe 20.6.

20.6 Desolder Components

Problem

You made a mistake and need to remove a soldered component from a board.

Solution

Desoldering is generally a lot harder than soldering. Unless the component you want to desolder is valuable, you can make your life easier by not worrying about destroying it to get it off the board.

For through-hole two-lead components like resistors, I use the following procedure:

1. Remove as much of the solder as possible by pressing the desoldering braid against the pad and then heating it with the soldering iron (Figure 20-22).
2. Sometimes this is so effective that you can then just pull out the component, but more often than not you will need to snip the component in half, or cut the lead at one end if there is enough spare lead to get a hold of with a pair of pliers on the top side of the board.

3. One leg at a time, hold the lead on the top side of the circuit board with pliers, and pull the lead out while heating the solder joint from below.

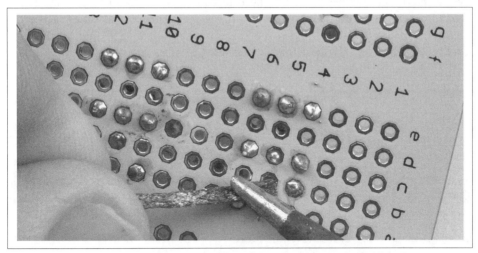

Figure 20-22. Using the Desoldering Braid

Desoldering DIL ICs is a lot more difficult, but again, if you don't mind sacrificing the IC, you can chop off its pins and desolder each one separately. Removing the IC as a whole is also possible. Starting with the desoldering braid, get the IC as clean of solder as possible and then push a screwdriver under one end of the IC and gently lever it off, swapping sides with the screwdriver to gradually work the IC out.

Desoldering SMDs is much easier. Just hold a hot-air gun over the component and it will probably just blow away after a while.

Discussion

Desoldering components is time consuming and you often end up damaging the circuit board pads or the component being removed. Sometimes it's quicker just to throw it in the trash and start again.

See Also

To solder through-hole devices, see Recipe 20.4 and for SMDs Recipe 20.5.

20.7 Solder Without Destroying Components

Problem

You have a high-power component (say a power transistor) and you want to know how big a heatsink to add to it to prevent it from overheating.

Solution

Determine the power the device will need to continuously dissipate and decide the maximum temperature (Tmax) you want the device to reach (less than its absolute maximum in the datasheet). Then use the following formula to find the thermal resistance the heatsink (RØheatsink) will need to provide:

$$R\theta_{heatsink} = \frac{T_{max} - T_{ambient}}{P} - R\theta_{package}$$

Here, Tambiant is the ambient temperature and RØpackage is the thermal resistance of the component package (1.5° C/W for a TO220 power transistor).

As an example, the datasheet for a TIP120 states that the maximum temperature is 150°C so let's stay well under this with a Tmax of 130°C. Let's assume the TIP120 is expected to be used in a circuit where it will generate 10W of heat.

Let's also assume the ambient temperature of the project's enclosure is 30°C (very dependent on ventilation).

Plugging these numbers into the formula, we have:

$$R\theta_{heatsink} = \frac{T_{max} - T_{ambient}}{P} - R\theta_{package} = \frac{130 - 30}{10} - 1.5 = 8.5 C/W$$

Searching through the heatsink listings in a component catalog, you can find heat-sinks with thermal resistances of 8.5C/W or better.

Discussion

Having done the calculations, test the temperature rise of the device for real and in its final enclosure, as the degree of ventilation in the enclosure will make a big difference. If need be, a fan on the heatsink will increase the heat dissipation considerably.

Heatsinks are available in a great variety of shapes and sizes. Figure 20-23 shows a selection of them.

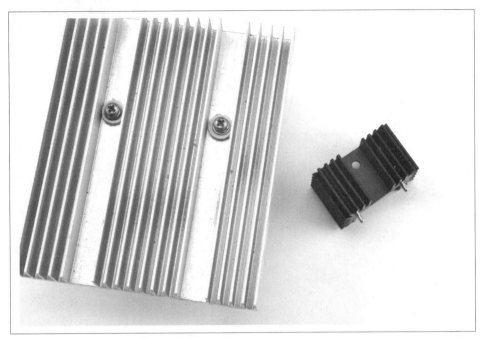

Figure 20-23. Heatsinks (Large and Small)

When attaching a device to a heatsink, you should smear a thin layer of heatsink compound (white paste) onto the device to greatly improve the heat transfer from the device to the heatsink.

See Also

For background on power, see Recipe 1.6.

The datasheet for the TIP120 is found here: *http://bit.ly/2mHBQy6*.

## CHAPTER 21

Tools

21.0 Introduction

This chapter explains how to use some of the most common electronic tools and test equipment. This includes both instruments for measurement such as multimeters and oscilloscopes and also simulation software that can be useful during design, especially when dealing with analog electronics.

21.1 Use a Lab Power Supply

Problem

You want to know how to correctly use a lab power supply.

Solution

To correctly use a lab power supply, follow these steps:

1. Set the voltage to the voltage your circuit needs.
2. Set the current limiting to slightly higher than the expected current consumption of your circuit.
3. Turn on the output and watch the voltage display. If too much current is being drawn, the voltage will drop, indicating that something is wrong.

Discussion

Aside from a multimeter, the lab power supply is probably the most useful piece of test equipment you can own. Owning one will save you vast amounts of time in the long run because you won't have to hunt around for batteries and cobble together power supplies and reduce the chance of accidental destruction of components when prototyping.

Figure 21-1 shows a typical lab power supply capable of supplying up to 22V at 5A.

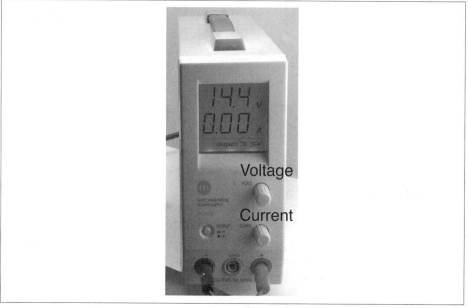

Figure 21-1. A Lab Power Supply

The top line of the display shows the voltage and the bottom, the current. When the output is off, the voltage knob allows you to set the voltage. The current knob sets the maximum current the circuit will be allowed to draw. If the circuit exceeds this, the power supply will automatically decrease the voltage until the current is below the set maximum. In this way, the power supply can be used as:

- Constant voltage with a maximum current
- Constant current with a maximum voltage

In addition to a single-output power supply like the one shown in Figure 21-1 dual-output power supplies are also available. These are very useful for analog circuits that require a split-voltage supply.

See Also

For information on building your power supplies of various sorts, see Chapter 7.

21.2 Measure DC Voltage

Problem

You want to measure DC voltage.

Solution

If you have an autoranging multimeter, simply set the meter to DC voltage and connect the probes across the voltage source.

If you have a multimeter where the range is set manually, determine the maximum voltage you expect to see and set the meter to the range whose maximum is above that value. Then connect the probe leads across the point in the circuit you want to measure.

Once you have established that the voltage is not too high for the range you selected, you can reduce the voltage range for better precision.

Discussion

Figure 21-2 shows a typical medium-range digital multimeter (DMM).

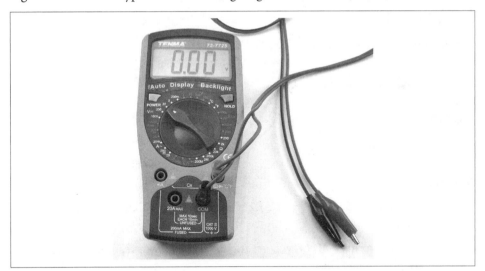

Figure 21-2. A Digital Multimeter

Although an autoranging multimeter might seem to be superior to a multimeter where the range has to be set, in practice, it can be an advantage to think about what reading you are expecting to see before you take the reading (manually setting the range forces you to do this).

Even very low-cost DMMs generally provide better precision and accuracy than an analog meter. The principle advantage of an analog meter is that you may get a few more clues about what you are measuring if, for example, the tip of the meter's needle jitters slightly to indicate noise, or you can see the rate of change of voltage as the speed of the meter moving in one direction or another. Some DMMs try to give you the best of both worlds by including an "analog" bargraph-type display in addition to the digital display.

Test Leads

Multimeters are generally supplied with probe-type test leads, which are fine for say measuring the voltage across a component, but often it is more convenient to be able to clip the leads in place while taking measurements.

It's also often quite handy to clip on the negative lead of the multimeter to ground and then use the probe end of the positive lead to measure the voltage at various parts of the circuit you are testing.

So, I highly recommend you buy yourself some multimeter leads that terminate in aligator clips, like the ones shown in Figure 21-2, to complement the regular test leads.

See Also

For measuring AC voltage, see Recipe 21.3.

21.3 Measure AC Voltage

Problem

You want to measure AC voltage.

Solution

Follow the same procedure as Recipe 21.2 except set the multimeter range to AC volts rather than DC volts.

Since you are measuring AC you will get the same polarity of reading no matter how the leads are placed onto the circuit.

If you are planning to measure high-voltage AC ensure your meter probes are rated for high voltage. Also see Recipe 21.12.

Discussion

Most DMMs will provide only an approximation of the RMS (root mean square) voltage by rectifying and smoothing it. Higher-end multimeters often include the feature "true RMS."

See Also

For measuring DC voltages, see Recipe 21.2.

21.4 Measure Current

Problem

You want to measure the current flowing through a certain point in a circuit.

Solution

To use a multimeter to measure current:

- Set the meter's range appropriately for AC/DC and at a higher range than the maximum current you expect.
- Fit the meter probes into the sockets indicated for current measurement. Note that these will be different sockets than for voltage measurement and there may be different sockets for different current ranges.
- Connect the leads to your circuit so the multimeter is in the current path. Figure 21-3 shows a DMM being used this way.

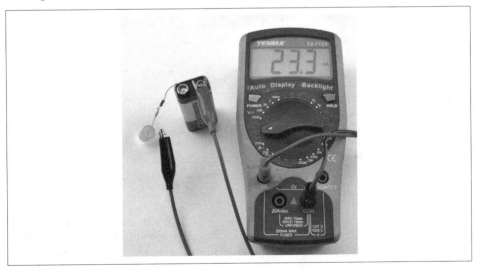

Figure 21-3. A Digital Multimeter Measuring Current

Discussion

DMMs measure the current flowing by measuring the voltage across a very low-value resistor. This is why you generally have to shift the probe leads to a different socket when measuring current.

Don't Forget to Swap the Leads Back

If you leave the leads of the multimeter in their current measuring sockets and then go to measure voltage, you will effectively short-out the voltage you are trying to measure. This may damage the circuit or blow a fuse inside the multimeter.

To prevent this, always swap the leads back to their voltage-measuring positions when you are finished measuring current.

If you do blow the fuse in your multimeter, you should be able to open up your multimeter's case and change the fuse.

See Also

To measure voltage, DC, and AC, see Recipe 21.2 and Recipe 21.3, respectively.

Most bench power supplies (Recipe 21.1) will also include an ammeter to tell you how much current is being drawn.

21.5 Measure Continuity

Problem

You have a wire, a copper track, or a fuse and you can't see a break in the connection but you want to test it electrically.

Solution

Disconnect the wire so that it is not in use and then use the continuity setting on your multimeter and connect the probes to each end of the thing you want to test.

If you are testing a long multicore cable (i.e., in situ and too long for the multimeter leads to reach both ends), you can connect the separate cores together at one end of the cable and test for continuity at the other, as shown in Figure 21-4.

Discussion

After DC voltage, continuity is probably the setting you are most likely to use on your multimeter. It is particularly useful if your multimeter makes a beeping noise when the resistance is low enough to indicate continuity. This allows you to move the test probes around without having to look at the multimeter's screen.

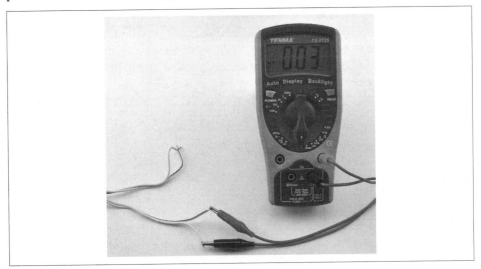

Figure 21-4. Testing a Long Cable for Continuity

See Also

Recipe 21.2 provides an introduction to multimeters.

21.6 Measure Resistance, Capacitance, or Inductance

Problem

You want to measure resistance, capacitance, or inductance (L) using a multimeter.

Solution

Nearly all multimeters will have several resistance ranges, and many will have a few capacitance ranges as well.

To use these ranges, simply select the range and attach the component to the test leads. You may find that as with measuring current, the test leads have to be inserted into different sockets on the multimeter when using these measurement ranges.

Discussion

Some DMMs will offer inductance and frequency ranges, and specialized meters are available for measuring resistance, capacitance, and inductance more accurately than the average DMM can.

Some of these will actually allow you to just attach test leads to any component and the meter will first identify it and then measure its properties. Amazingly, such meters are available on eBay as a kit to assemble yourself for around $10.

When measuring component values, do not be seduced into thinking you are making a very accurate measurement of a component value because of the precision of the result. A reading of 1.23μF may still have a measurement of ±10%, so check the specification for your meter.

See Also

Recipe 21.2 provides an introduction to multimeters.

21.7 Discharge Capacitors

Problem

You have a large capacitor that could be holding a lot of energy and you want to safely discharge it.

Solution

Disconnect the circuit and then use a resistor in parallel with the capacitor to discharge the capacitor until the voltage across it has reached a safe level as indicated by a multimeter set to its DC-voltage setting.

When discharging the capacitor, you can either connect the resistor using insulated crocodile clips, or if access is good enough, bend the resistor leads to the correct gap and then hold the resistor with pliers, carefully touching the capacitor leads as shown in Figure 21-5.

Calculate the resistance and the power rating of the resistor such that the capacitor discharges in a reasonable amount of time without the resistor becoming too hot.

The time constant (RC) is the time in seconds for the capacitor's voltage to drop to 63.2% of its original value. For example, if you have a 100μF capacitor charged to 300V (you might find this in a photographic flashgun) and you want to discharge the capacitor to a safe 10V, a resistor of 10kΩ would have a time constant of 1 second. So holding it to the capacitor for 1 second would reduce the voltage to 190V, a further second to 120V, and so on. So, after 7 seconds, the voltage would be down to a safe 7.6V.

You can calculate the maximum power the resistor will dissipate as heat using:

$$P = \frac{V^2}{R}$$

Figure 21-5. Discharging a Capacitor

which in this case would be 9W. That's quite a physically big resistor, so failing to do a rough calculation and using a standard ¼W resistor is likely to end in a puff of smoke for the resistor.

Discussion

Increasing the resistance of the resistor decreases the power requirement of the resistor but will take longer for the capacitor to discharge. It is a good idea to monitor the capacitor's voltage with a multimeter while you discharge.

A capacitor charged to a high voltage is a dangerous thing, but a high-value capacitor even at low voltages can cause massive currents to flow if its terminals are short circuited and the capacitor has a low ESR.

See Also

For a calculation on the energy stored in a capacitor, see Recipe 3.7.

21.8 Measure High Voltages

Problem

You want to measure a voltage that is higher than the maximum voltage range of your multimeter.

Solution

Use a voltage divider comprised of a ladder of equal-value resistors to reduce the voltage to be measured. You will need to take into account the effect that the voltage divider has on the voltage being measured and the input impedance of the multimeter. Also, make sure the resistors used are rated for a high enough voltage.

Figure 21-6 shows how you might measure a voltage of around 5kV using a multimeter with a maximum DC voltage measurement of 1000V and an imput impedance of 10MΩ.

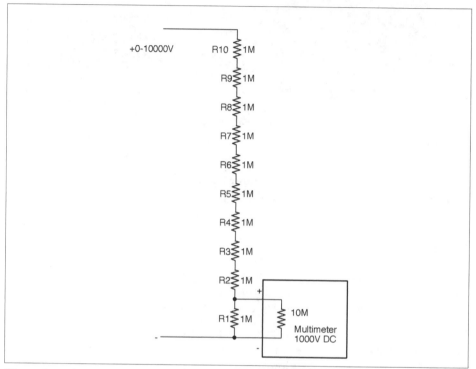

Figure 21-6. Measuring a High Voltage Using a Voltage Divider

The voltage divider will reduce the input voltage by a factor of 10, making the sums easier. Using 10 equal-value resistors (of 1% or better accuracy) is more likely to result in better accuracy of the system overall if the resistors are from the same batch. The closer the resistors are in value to each other the better the divider.

There are a few other things you need to consider. First, remember to calculate how much the chain of resistors will load the output of the high-voltage source. In this case, a load of 10MΩ across 10kV will result in current flowing at 1mA.

High-Voltage Leads

If you plan to measure high voltages, you will need special high-voltage leads that are extremely well insulated to prevent sparking across to your fingers. Your voltage divider should also be boxed and not easy to accidentally touch.

Also, see Recipe 21.12.

The heat power generated by each resistor will be 1kV * 1mA = 1W, which is significant. The temptation is to use higher-value resistors (say 10MΩ) to reduce the power and loading on the voltage source, but this will lead to the impedance of the multimeter becoming significant and acting as two similar-value resistors in parallel, making the reading almost useless.

A typical low-cost or medium-range multimeter will only have an input impedance of 10MΩ, which would result in the voltage reading being reduced by about 10% from the actual voltage.

Discussion

You can determine the input impedance of your multimeter using the schematic in Figure 21-7.

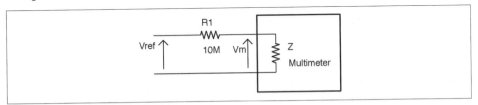

Figure 21-7. Finding the Input Impedance of Your Multimeter

To determine the input impedance of your multimeter (Z in Figure 21-7):

1. Accurately measure the voltage of a stable voltage source—say a regulated 5V supply—by directly connecting the multimeter leads across it. Make a note of this value (Vref).
2. Now place a resistor in series with the positive multimeter leads as shown in Figure 21-7 and see what voltage the meter reads now (Vm). If there is little or no change in the reading, then try a higher-value resistor (say 100MΩ) and congratulations you have a high-quality multimeter.
3. Calculate the impedance of the meter (Z) using the formula:

$$Z = \frac{R}{\left(\frac{V_{in}}{V_{out}} - 1\right)}$$

For example, with a value of R1 of 10MΩ and a 10V supply, my meter gave a Vm reading of 4.7V. Plugging the numbers in you get:

$$Z = \frac{R}{\left(\frac{V_{in}}{V_{out}} - 1\right)} = \frac{10M}{\left(\frac{10}{4.7} - 1\right)} = \frac{10M}{1.13} = 8.87M\Omega$$

If you plan to measure high voltages on a regular basis, buy a specialist high-voltage voltmeter. In addition to having a very high voltage range, these instruments also usually have very high input impedance and do not load the circuit under test to the extent that the voltage being measured is appreciably altered.

You can also use the preceding method using very high-value resistors if you buy a high-quality multimeter with a buffered input that gives the multimeter an input impedance in the hundreds of MΩ or even GΩ range.

See Also

For more information on using a resistor as a voltage divider, see Recipe 2.6.

For regular DC-voltage measurement, see Recipe 21.2.

21.9 Use an Oscilloscope

Problem

You want to use an oscilloscope to see the waveform of a signal.

Solution

Figure 21-8 shows a typical low-cost oscilloscope displaying a 1kHz 5V test signal available from a terminal on the front of the oscilloscope.

To look at a signal on an oscilloscope:

1. Estimate the maximum signal voltage and set the y-axis gain to a value that will allow you to see the whole signal on the screen. For example, in this case the signal is 5V so setting the y-axis gain for the channel being used (channel A) to 2V/division (little square on the screen) will be fine. If in doubt, set the volts/division to its maximum value.
2. Set the trigger level to a height up the screen where the changing signal will cause the scope to trigger, allowing you to see a stable image of the signal.
3. Adjust the x-axis timebase until the signal is stretched out enough to see the shape of the waveform. In this case, the x-axis timebase is set to 500µS/division so one whole waveform occupies two divisions (1ms), confirming the frequency of 1kHz.

Discussion

Every model of oscilloscope is a little different, so you will need to consult the manual for yours to find the controls used in the preceding instructions.

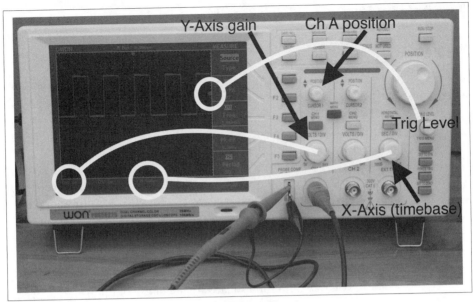

Figure 21-8. Displaying a Signal on an Oscilloscope

Most oscilloscopes, including the one shown in Figure 21-8, have two channels that allow you to display two signals at the same time as well as a host of other features such as automatic measurement of frequency and signal amplitude.

When choosing an oscilloscope to buy, you can spend from a few hundred dollars to many thousands of dollars. You pay more for higher frequency range, better displays, and more advanced features. A basic 20MHz oscilloscope like the one shown in Figure 21-8 is a perfectly good starting point and has served me well for many years.

In addition to standalone oscilloscopes, you can also buy "PC scopes" that do not have a screen but rather rely on a USB connection to the PC running the oscilloscope software. As with standalone oscilloscopes, PC scopes are available at all prices and qualities. I prefer a standalone device as it's always there on my workbench and I don't have to wait for it to boot up, but many people find the extra features that often come with a PC scope and the bigger and better display of a computer monitor outweigh any disadvantages.

See Also

For more information on getting the most from your oscilloscope, get to know the manual really thoroughly. You may find all sorts of features that are not immediately obvious from the controls on the front.

21.10 Use a Function Generator

Problem

You need a test signal at a certain frequency, a certain amplitude, DC offset, and waveform shape, perhaps to test an amplifier or filter.

Solution

Use a function generator (a.k.a. a signal generator).

The function generator shown in Figure 21-9 is a typical low-cost function generator capable of generating two independent sine, square, or triangle waveforms of up to 20 MHz.

To use the function generator:

- Turn its output off
- Set the waveform you want (sine and square are the most common)
- Set the p-p (peak-to-peak) amplitude
- Set the DC offset
- Optionally attach one channel of an oscilloscope to the signal and test it appears as expected by turning on the signal generator's output for a moment
- Connect the signal-generator output to the input of the circuit you are testing
- Turn the signal-generator output on

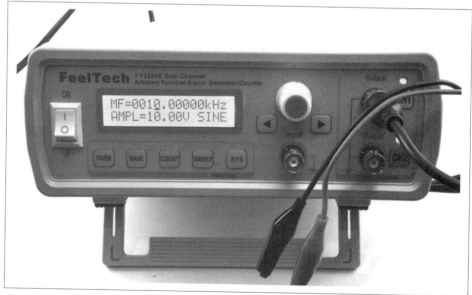

Figure 21-9. A Low-cost DDS (Direct Digital Synthesis) Function Generator

Don't Forget DC Offset

If you have a single-supply amplifier or other circuit, if the input signal is swinging negative on each cycle, you may damage the circuit you are testing.

Signal generators with digital controls, like the one shown in Figure 21-9, assume a full AC signal and a DC offset have to be explicitly set.

Discussion

Figure 21-10 shows the oscilloscope trace for a 10kHz sinewave with an amplitude of 2V peak-to-peak and a DC offset of 2.5V generated by the function.

See Also

If you are on a budget, you can make your own oscillator as shown in Recipe 16.5.

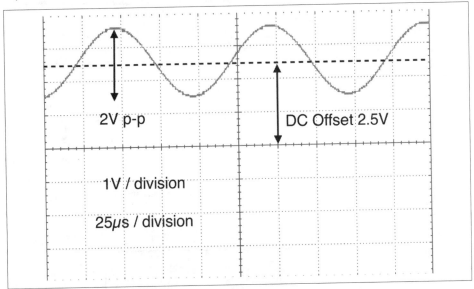

Figure 21-10. Waveform Amplitude and DC Offset

21.11 Simulation

Problem

You want to be able to simulate how a circuit will behave before you make it, perhaps simulating how well a filter will work.

Solution

Use circuit-simulator software.

A free online circuit simulator is a great way to get started with circuit simulation. The PartSim (*http://partsim.com*) is one such easy-to-use simulator. Sign up for a PartSim account and then start drawing your schematic in the simulator. Figure 21-11 shows the schematic for the simple RC filter from Recipe 16.3.

In addition to drawing R1 and C1, you can also specify an AC-voltage source to drive it. In this case, as the parameter in Figure 21-11 suggests, the test signal will be a 5V square wave (pulse) with rise and fall times of 1μs, a pulse length of 15μs, and an overall period of 30μs. This corresponds roughly to the 32.7kHz carrier signal of Recipe 16.3.

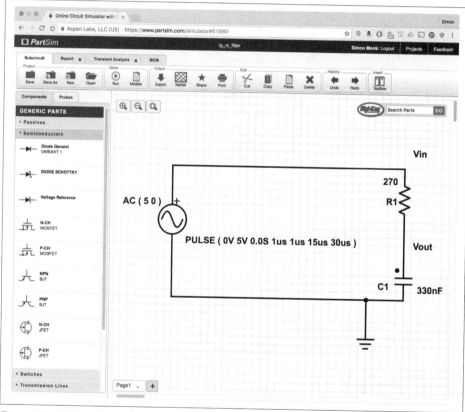

Figure 21-11. PartSim Schematic Editor

When you click the Run button you will be prompted to enter some parameters for a simulation as shown in Figure 21-12.

There are various types of simulations that can be run, but in this case we are interested in the "Transient Response." The Start and Stop times determine how long the simulation will run and the Time Step is the step between each calculated value in the simulation.

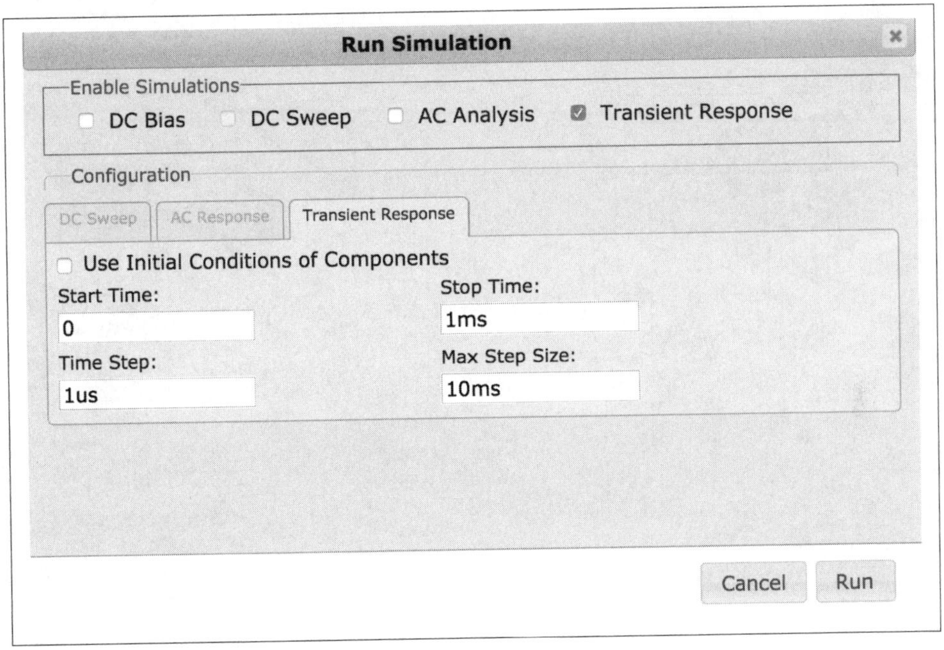

Run Simulation ✕

┌─Enable Simulations──┐
│ ☐ DC Bias   ☐ DC Sweep   ☐ AC Analysis   ☑ Transient Response │
└──┘

┌─Configuration───┐
│ DC Sweep │ AC Response │ **Transient Response** │
│   │
│ ☐ Use Initial Conditions of Components   │
│ Start Time: Stop Time: │
│ 0   1ms   │
│ Time Step: Max Step Size: │
│ 1us   10ms   │

Cancel   Run

Figure 21-12. Simulation Parameters

When you click Run, you will see a new tab appear in the window with the name "transient response" that shows the result of the simulation as shown in Figure 21-13.

You can see that the output is indeed greatly attenuated by the RC filter.

Discussion

Simulation is very useful in analog design, not least because it tells you how the circuit *should* behave, whereas a signal generator and oscilloscope will tell you how just one prototype behaves. A physical prototype has the problem that there may be a fault in its construction or components that causes it to behave differently when you make a second prototype. Simulation will tell you what to expect in a reliable and consistent manner.

In addition to idealized components like resistors, capacitors, and perfect op-amps, simulators like PartSim also have a huge array of "models" for actual components, including specific op-amp models.

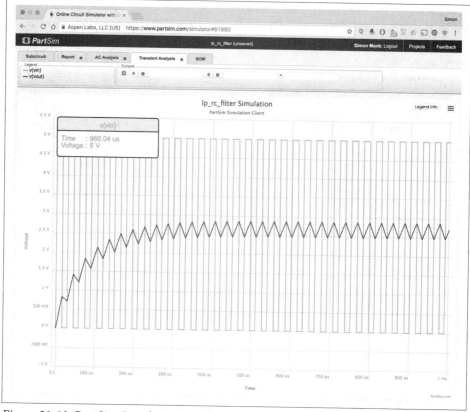

Figure 21-13. PartSim Simulation Results

See Also

PartSim is based on the open source SPICE simulation software; learn more at: *http://bwrcs.eecs.berkeley.edu/Classes/IcBook/SPICE/*.

21.12 Working Safely with High Voltages

Problem

You want to work safely with high voltages and avoid electrocution, burns, and fires.

Solution

Assume that coming into contact with a high voltage will kill you.

Although a little hyperbolic, this statement is a good thing to keep at the front of your mind when working with high voltages. Frankly anything above 50V should scare you. So AC with its fatal combination of high voltage and high availability of current should scare you a lot.

Here are some rules I stick to when working on AC:

- If you are not sure you have the skills and experience to work on AC, find a friend to help you who does, or do something else.
- Never work on a project while it is connected to AC. I actually put the plug in front of me, so that I can see it's not plugged in. Don't rely on the outlet switch.
- Don't do this kind of work if you feel tired. That's when mistakes happen.
- Always discharge any capacitors in the project (see Recipe 21.7).
- If you make a project using high voltages, always make sure it is enclosed so that other people cannot accidentally get shocked.
- Connect any metal of your project to earth.
- Use standard connectors (such as "kettle leads") that are correctly rated for the voltage and current you are using.
- Think about what you are doing and check everything before you power up.

Discussion

According to the American Burn Association:

In the United States, on average of 400 people die from electrocution and 4,400 are injured each year because of electrical hazards.

In addition to the risk of a current flowing through your heart and stopping it, burns caused by your body effectively acting as a heating element as well as arcing sparks are risks associated with high voltages.

See Also

For the full American Burn Association report, see: *http://www.ameriburn.org/ Preven/ElectricalSafetyEducator'sGuide.pdf.*

Parts and Suppliers

Parts

The following tables will help you to find the parts used in this book. Where possible, I have listed product codes for suppliers.

There are now many electronic component suppliers that cater to the maker and electronics hobbyist. Some of the most popular are listed in Table A-1.

Table A-1. Parts Suppliers

Supplier	Website	Notes
Adafruit	*http://www.adafruit.com*	Good for modules
DigiKey	*http://www.digikey.com/*	Wide range of components
MakerShed	*http://www.makershed.com/*	Good for modules, kits, and tools
MCM Electronics	*http://www.mcmelectronics.com/*	Wide range of components
Mouser	*http://www.mouser.com*	Wide range of components
SeeedStudio	*http://www.seeedstudio.com/*	Interesting low-cost modules
SparkFun	*http://www.sparkfun.com*	Good for modules
MonkMakes	http://www.monkmakes.com	Electronic Kits for Raspberry Pi, etc.
Pimoroni	https://shop.pimoroni.com	Raspberry Pi and Arduino
Polulu	https://www.pololu.com/	Great for motor controllers and robots
CPC	*http://cpc.farnell.com/*	UK-based, wide range of components
Farnell	*http://www.farnell.com/*	International, wide range of components
Maplin	*http://www.maplin.co.uk/*	UK-based, bricks and mortar, Raspberry Pi and Arduino
Proto-pic	*http://proto-pic.co.uk/*	UK-based, stock SparkFun, and Adafruit modules

The other great source for components is eBay.

Searching for components can be time consuming and difficult. The Octopart component search engine (*http://www.octopart.com*) can be very helpful for tracking down parts.

Prototyping Equipment

Many of the hardware projects in this book use jumper wires of various sorts. Male-to-female leads (to connect the Raspberry Pi GPIO connector to a breadboard) and male-to-male (to make connections on the breadboard) are particularly useful. Female-to-female are occasionally useful for connecting modules directly to GPIO pins. You rarely need leads longer than 3 inches (75mm). Table A-2 lists some jumper wire and breadboard specifications, along with their suppliers.

A handy way to get started with breadboard, jumper wires, and some components to get you started is to buy a starter kit like the Hacking Electronics Kit or Electronics Starter Kit for Raspberry Pi by MonkMakes.com (*http://monkmakes.com*).

Table A-2. Prototyping Equipment

Description	Suppliers
M-M jumper wires	SparkFun: PRT-08431, Adafruit: 759, DigiKey: PRT-08431-ND
M-F jumper wires	SparkFun: PRT-09140, Adafruit: 825, DigiKey: PRT-09140-ND
F-F jumper wires	SparkFun: PRT-08430, Adafruit: 794, DigiKey: PRT-08430-ND
Half-sized breadboard	SparkFun: PRT-09567 Adafruit: 64, DigiKey: 377-2094-ND
Raspberry Leaf (26 pin)	Adafruit: 1772
Raspberry Leaf (40 pin)	Adafruit: 2196
Electronics Starter Kit for Raspberry Pi	Amazon, monkmakes.com
Monk Makes Protoboard	Amazon, monkmakes.com/pb
Adafruit PermaProto for Pi (half breadboard)	Adafruit: 1148
Adafruit PermaProto for Pi (full breadboard)	Adafruit: 1135
Adafruit PermaProto HAT	Adafruit: 2314, DigiKey: 1528-1370-ND
DC barrel jack to screw terminal adapter (female)	Adafruit: 368, DigiKey: 1528-1386-ND

Resistors

Table A-3 lists resistors used in this cookbook and some suppliers.

Table A-3. Resistors

10Ω 0.25W resistor	Mouser: 293-10-RC, DigiKey: 10QBK-ND
22Ω 0.25W resistor	Mouser: 293-22-RC, DigiKey: 22QBK-ND
100Ω 0.25W resistor	Mouser: 293-100-RC, DigiKey: 100QBK-ND
120Ω 0.25W resistor	Mouser: 293-120-RC, DigiKey: 120QBK-ND

150Ω 0.25W resistor	Mouser: 293-150-RC, DigiKey: 150QBK-ND
270Ω 0.25W resistor	Mouser: 293-270-RC, DigiKey: 270QBK-ND
330Ω 0.25W resistor	Mouser: 293-330-RC, DigiKey: 330QBK-ND
470Ω 0.25W resistor	Mouser: 293-470-RC, DigiKey: 470QBK-ND
1kΩ 0.25W resistor	Mouser: 293-1k-RC, DigiKey: 1.0kQBK-ND
3.3kΩ 0.25W resistor	Mouser: 293-3.3k-RC, DigiKey: 3.3kQBK-ND
4.7kΩ 0.25W resistor	Mouser: 293-4.7k-RC, DigiKey: 4.7kQBK-ND
10kΩ 0.25W resistor	Mouser: 293-10k-RC, DigiKey: 10kQBK-ND
22kΩ 0.25W resistor	Mouser: 293-22k-RC, DigiKey: 22kQBK-ND
33kΩ 0.25W resistor	Mouser: 293-33k-RC, DigiKey: 33kQBK-ND
100kΩ 0.25W resistor	Mouser: 293-100k-RC, DigiKey: 100kQBK-ND
180kΩ 0.25W resistor	Mouser: 293-180k-RC, DigiKey: 180kQBK-ND
1MΩ 0.25W resistor	Mouser: 293-1M-RC, DigiKey: 1.0MQBK-ND
1.8MΩ 0.25W resistor	Mouser: 293-1.8M-RC, DigiKey: 1.8MQBK-ND
10 kΩ trimpot	Adafruit: 356, SparkFun: COM-09806, Mouser: 652-3362F-1-103LF, DigiKey: 3386P-103TLF-ND
Photoresistor	Adafruit: 161, SparkFun: SEN-09088, DigiKey: NSL-5152-ND
Thermistor T0 of 1k Beta 3800 NTC	Mouser: 871-B57164K102J (note Beta is 3730), DigiKey: 495-75312-ND

Capacitors and Inductors

Table A-4 lists resistors and capacitors used in this cookbook and some suppliers.

Table A-4. Resistors and Capacitors

1nF 50V	DigiKey: BC2659CT-ND, Mouser: 594-K102J15C0GF5TH5
10nF 50V	DigiKey: BC2662CT-ND, Mouser: 594-K103K15X7RF5UL2
10nF 1000V	DigiKey: 1255PH-ND, Mouser: 81-RDER73A103K3M1H3A
100nF 50V	DigiKey: 399-4151-ND, Mouser: 594-K104K15X7RF53L2
100nF 400V	DigiKey: EF4104-ND, Mouser: 581-SR758C104KAATR1
220nF 50V	DigiKey: BC2678CT-ND, Mouser: 594-K224K20X7RF5TH5
330nF 50V	DigiKey: 399-9882-1-ND, Mouser: 80-C330C334K5R
680nF 50V	DigiKey: 445-8519-ND, Mouser: 81-RCER71H684K2M1H3A
1µF 16V	DigiKey: 445-8614-ND, Mouser: 539-SN010M025ST
4.7µF 16V	DigiKey: 493-10248-1-ND, Mouser: 647-UMA1C4R7MCD2
10µF 16V	DigiKey: 493-10245-1-ND, Mouser: 667-ECE-A1CKS100
100µF 16V	DigiKey: P16379CT-ND, Mouser: 598-107CKS016M
220µF 25V	DigiKey: 493-6082-ND, Mouser: 667-EEU-FM1E221
470µF 35V	DigiKey: 493-12724-1-ND, Mouser: 667-ECA-1VM471
1000µF 25V	DigiKey: 493-12690-1-ND, Mouser: 667-EEU-FC1E102L
390nH 100mA	DigiKey: 445-1010-1-ND, Mouser: 542-9230-10-RC
4.7µH 250mA	DigiKey: 495-5567-1-ND, Mouser: 70-IR04RU4R7K

22uH 3A Inductor   DigiKey: 495-5590-1-ND, Mouser: 580-12RS223C

33uH 3A Inductor   DigiKey: 495-5705-1-ND, Mouser: 963-LHL13NB330K

Transistors, Diodes

Table A-5 lists transistors and diodes used in this cookbook and some suppliers.

Table A-5. Transistors and Diodes

FQP30N06L N-channel logic-level MOSFET transistor	Mouser: 512-FQP30N06L, SparkFun: COM-10213, DigiKey: FQP30N06L-ND
FQP27P06 P-channel MOSFET transistor	SparkFun: COM-10349, Mouser: 512-FQP27P06, DigiKey: FQP27P06-ND
2N3904 NPN bipolar transistor	SparkFun: COM-00521, Adafruit: 756, Mouser: 512-2N3904BU, DigiKey: 2N3904TAFSCT-ND
2N3906 PNP bipolar transistor	SparkFun: COM-00522, Mouser: 512-2N3906TA, DigiKey: 2N3906-APCT-ND
TIP120 Darlington transistor	Adafruit: 976, CPC: SC10999, Mouser: 511-TIP120, DigiKey: TIP120-ND
2N7000 MOSFET transistor	Mouser: 512-2N7000, CPC: SC06951, DigiKey: 2N7000TACT-ND
STGF3NC120HD IGBT	Mouser: 511-STGF3NC120HD, DigiKey: 497-4353-5-ND
IRG4PC30UPBF IGBT	Mouser: 942-IRG4PC30UPBF
1N4001 diode	Mouser: 512-1N4001, SparkFun: COM-08589, Adafruit: 755, DigiKey: 1N4001DICT-ND
1N4004 diode	Mouser: 512-1N4004, DigiKey: 1N4004FSCT-ND
1N4007 diode (1000V)	Mouser: 821-1N4007, DigiKey: 1N4007FSCT-ND
1N4148 diode	Mouser: 512-1N4148, DigiKey: 1N4148FS-ND
1N5819 Schottky diode	Mouser: 512-1N5819, DigiKey: 1N5819FSCT-ND
1N5919 5.6V Zener diode	Mouser: 863-1N5919BG, DigiKey: 1N5919BGOS-ND
BT136 TRIAC	Mouser: 583-BT136, DigiKey: 568-12097-5-ND

Figure A-1 shows the pinouts for the transistors listed in this section.

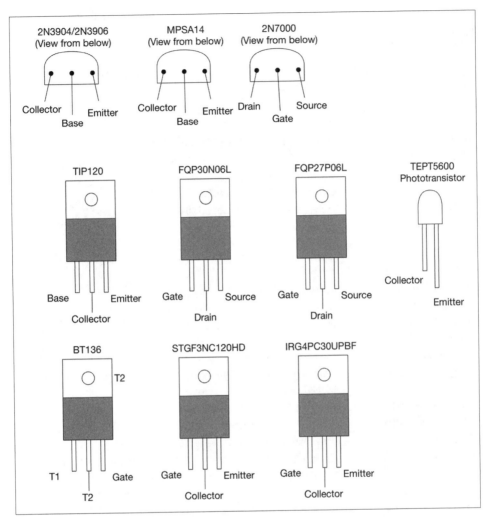

Figure A-1. Transistor Pinouts

Integrated Circuits

Table A-6 lists ICs used throughout this cookbook and some suppliers. These are listed in alphabetical order by part name.

Table A-6. Integrated Circuits

74HC00 quad NAND	DigiKey: 296-1563-5-ND, Mouser: 595-SN74HC00N
74HC4017 counter decoder	DigiKey: 296-25989-5-ND, Mouser: 595-CD74HC4017E
74HC4094 shift register	DigiKey: 296-26002-5-ND , Mouser: 595-CD74HC4094E
74HC590 counter	DigiKey: 296-1599-5-ND, Mouser: 595-SN74HC590AN
CD4047 oscillator	DigiKey: 296-2053-5-ND, Mouser: 595-CD4047BEE4
DS18B20 temperature sensor	SparkFun: SEN-00245, Adafruit: 374, Mouser: 700-DS18B20, CPC: SC10426, DigiKey: DS18B20+-ND
L293D motor driver	SparkFun: COM-00315, Adafruit: 807, Mouser: 511-L293D, CPC: SC10241, DigiKey: 497-2936-5-ND
LM2596-5V switcher	DigiKey: LM2596T-5.0/NOPB-ND, Mouser:
LM311 comparator	DigiKey: 296-1389-5-ND, Mouser: 926-LM311N/NOPB
LM317 adjustable voltage regulator	DigiKey: LM317AHVT-ND , Mouser: 595-LM317KCSE3
LM321 op-amp	DigiKey: LM321MFX/NOPBCT-ND, Mouser: 926-LM321MF/NOPB
LM741 op-amp	DigiKey: LM741CNNS/NOPB-ND, Mouser: 926-LM741CN/NOPB
LM7805 voltage regulator	SparkFun: COM-00107, Adafruit: 2164, Mouser: 511-L7805CV, CPC: SC10586, DigiKey: 497-1443-5-ND
LM78L12 voltage regulator	DigiKey: LM78L12ACZFS-ND, Mouser: 512-LM78L12ACZ
LM79L12 voltage regulator	DigiKey: LM79L12ACZ/NOPB-ND, Mouser: 926-LM79L12ACZ/NOPB
MAX2606 VCO	DigiKey: MAX2606EUT+TCT-ND , Mouser: 700-MAX2606EUTT
MCP3008 eight-channel ADC IC	Adafruit: 856, Mouser: 579-MCP3008-I/P, CPC: SC12789, DigiKey: MCP3008-I/P-ND
MCP73831 LiPo charger IC	DigiKey: MCP73831T-2DCI/OTCT-ND, Mouser: 579-MCP73831T5ACI0T
NE555 timer	SparkFun: COM-09273, DigiKey: 296-1411-5-ND, Mouser: 595-NE555P
OPA365 op-amp	DigiKey: 296-20645-1-ND, Mouser: 595-OPA365AIDBVR
TDA7052 1W power amp	DigiKey: 568-1138-5-ND, Mouser: 771-TDA7052ATN2112
TLV2770 op-amp	DigiKey: 296-1897-5-ND, Mouser: 595-TLV2770IP
TPA3122D2 15W power amp	DigiKey: 296-23375-5-ND , Mouser: 595-TPA3122D2N
TMP36 temperature sensor	SparkFun: SEN-10988, Adafruit: 165, Mouser: 584-TMP36GT9Z, CPC: SC10437, DigiKey: TMP36GT9Z-ND
TPS61070 boost converter	DigiKey: 296-17151-1-ND, Mouser: 595-TPS61070DDCR
ULN2803 Darlington driver IC	SparkFun: COM-00312, Adafruit: 970, Mouser: 511-ULN2803A, CPC: SC08607, DigiKey: 497-2356-5-ND
WS2812 pixel chip	DigiKey: 28085-ND
MOC3032 opto-isolator	DigiKey: MOC3032M-ND, Mouser: 512-MOC3032M

Figure A-2 shows the pinouts for the ICs listed in this section.

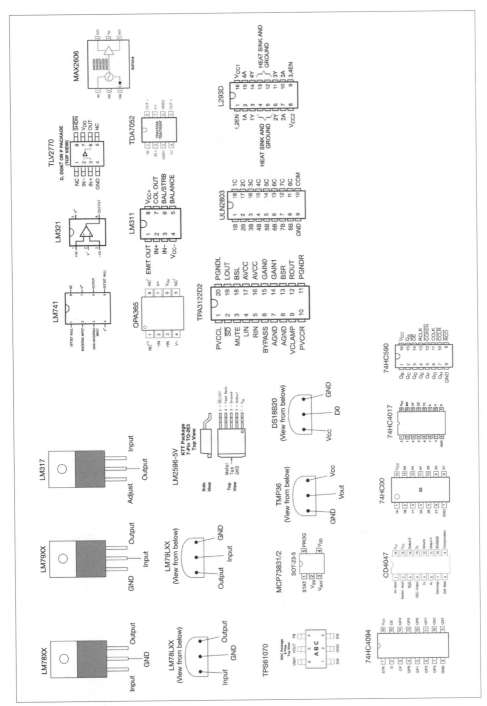

Figure A-2. IC Pinouts

Opto-Electronics

Table A-7 lists opto-electronic components used throughout this cookbook and some suppliers.

Table A-7. Opto-electronics

5mm red LED	SparkFun: COM-09590, Adafruit: 299, Mouser: 630-HLMP-3301, DigiKey: 160-1853-ND
RGB common cathode LED	SparkFun: COM-11120, Mouser: 713-104990023, eBay
TSOP38238 IR sensor	SparkFun: SEN-10266, Adafruit: 157
4-digit 7-segment common cathode LED display	DigiKey: 67-1450-ND

Modules

Table A-8 lists the modules used in this cookbook and other modules I like.

Table A-8. Modules

Arduino Uno	SparkFun: DEV-11021, Adafruit: 50, CPC: A000066, DigiKey: 1050-1024-ND
Raspberry Pi3	Adafruit: 3055, DigiKey: 1690-1000-ND
Level converter, four-way	SparkFun: BOB-11978, Adafruit: 757
Level converter, eight-way	Adafruit: 395
LiPo boost converter/charger	SparkFun: PRT-11231
PowerSwitch tail	Adafruit: 268
Monk Makes ServoSix board	monkmakes.com, Amazon
16-channel servocontroller	Adafruit: 815
Motor driver 1A dual	SparkFun: ROB-09457
RasPiRobot board V3	Adafruit: 1940, Amazon
PIR motion detector	Adafruit: 189
4x7-segment LED with I2C backpack	Adafruit: 878
Bicolor LED square-pixel matrix with I2C backpack	Adafruit: 902
16 x 2 HD44780 compatible LCD module	SparkFun: LCD-00255, Adafruit: 181
SSD1306-based 0.96 or 1.2in OLED display	eBay
Stepper motor HAT	Adafruit: 2348
16-channel PWM HAT	Adafruit: 2327
Squid Button	monkmakes.com, Amazon
Raspberry Squid RGB LED	monkmakes.com, Amazon
I2C OLED display 128x64 pixels	eBay
Adafruit Lipo charger module	Adafruit: 1905
SparkFun LiPo charger module	SparkFun: PRT-10217
CC1101 RF transceiver module	eBay

Miscellaneous

Table A-9 lists miscellaneous tools and components used in this cookbook and some suppliers.

Table A-9. Miscellaneous

1200mAh LiPo battery	Adafruit: 258
5V relay	SparkFun: COM-00100
Standard servomotor	SparkFun: ROB-09065, Adafruit: 1449
9g mini servomotor	Adafruit: 169
5V 1A power supply	Adafruit: 276
Low power 6V DC motor	Adafruit: 711
0.1-inch header pins	SparkFun: PRT-00116, Adafruit: 392
5V 5-pin unipolar stepper motor	Adafruit: 858
12V, 4-pin bipolar stepper motor	Adafruit: 324
Tactile push switch	SparkFun: COM-00097, Adafruit: 504
Miniature slide switch	SparkFun: COM-09609, Adafruit: 805
Rotary encoder (quadrature)	Adafruit: 377
4x3 keypad	SparkFun: COM-08653
Piezo buzzer	SparkFun: COM-07950, Adafruit: 160
Reed switch	Adafruit: 375
Loudspeaker 8Ω 1W	Adafruit: 1313

Equipment

There are many choices here. I would always say start with low-cost equipment and upgrade as and when you feel the need. After all, if you were learning the violin, it would be foolish to start with a Stradivarius!

The items listed in Table A-10 are similar to those I use every day and serve as a guide to help you get started. You should shop around; there are some great bargains to be had.

Table A-10. Equipment

Basic multimeter	Monk Makes Hacking Electronics Kit, eBay
Better multimeter (Tenma 72-7725)	Amazon, eBay
Entry-level oscilloscope	Adafruit: 681
Bench power supply	DigiKey: BK1550-220V-ND
Good soldering station	SparkFun: TOL-11704
Fume extractor	eBay
Thermal heatsink compound	eBay

Arduino Pinouts

Arduino Uno R3

Figure B-1 shows the pinout for the Arduino Uno R3.

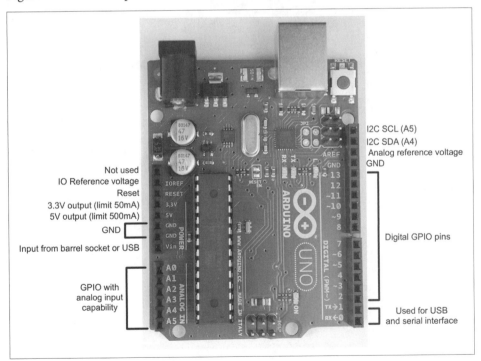

Figure B-1. Arduino Uno R3 GPIO Pinout

Arduino Pro Mini

Figure B-2 shows the pinout for the Arduino Pro Mini.

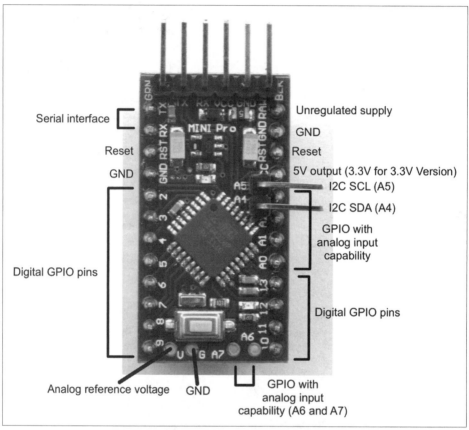

Figure B-2. Arduino Uno Pro Mini Pinout

Raspberry Pi Pinouts

Raspberry Pi 2 Model B, B+, A+, Zero

Figure C-1 shows the pinouts for the current 40-pin GPIO Raspberry Pi.

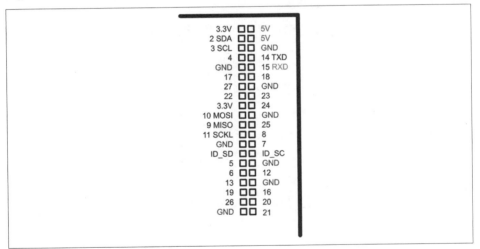

Figure C-1. 40-pin Raspberry Pi GPIO Pinout

Raspberry Pi Model B, Rev. 2, A

If you have an Raspberry Pi, it is most likely the model B rev. 2 board shown in Figure C-2.

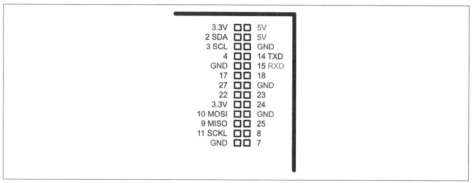

Figure C-2. Raspberry Pi Model B rev. 2 and Model A GPIO Pinout

Raspberry Pi Model B, Rev. 1

The very first released version of the Raspberry Pi model B (rev. 1) has some minor pinout differences from the rev. 2 that followed. This is the only version of the Raspberry Pi that is not compatible with later pinouts. The incompatible pins that changed are highlighted in bold in Figure C-3.

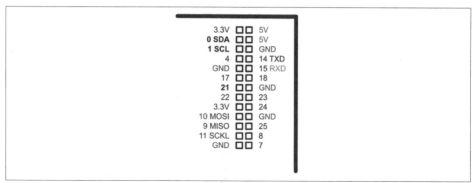

Figure C-3. Raspberry Pi Model B Rev. 1 GPIO Pinout

Units and Prefixes

Units

Table D-1 lists the most common units used in electronics along with the typical ranges of those units you might find in practice.

Table D-1. Common Units and their Typical Ranges of Values

Property	Unit	Typical range
Current	A (Ampere)	100nA to 100A
Voltage	V (Volt)	1mV to 1000V
Resistance	Ω (Ohm)	10mΩ to 20MΩ
Energy	J (Joule)	1J to 1MJ
Watt	W (Watt)	1mW to 10kW
Capacitance	F (Farad)	10pF to 10F
Inductance	H (Henry)	
Frequency	Hz (Hertz)	sound 20Hz to 20kHz radio 3kHz to 300GHz

Unit Prefixes

Table D-2 shows the unit prefixes applied to electronic and other units. You will sometimes find R used in place of Ω and u in place of μ.

Table D-2. Unit Prefixes

Prefix	Multiplier	Scientific
p (pico)	1/1,000,000,000,000	10E-12
n (nano)	1/1,000,000,000	10E-9

Prefix	Multiplier	Scientific
μ (micro)	1/1,000,000	10E-6
m (milli)	1/1,000	10E-3
k (kilo)	1,000	10E3
M (mega)	1,000,000	10E6
G (giga)	1,000,000,000	10E9

Index

Symbols

I

J

K

L

M

NPN (negative positive negative), 58
opto-couplers, 69, 416
oscillator creation using, 297
phototransistors, 68
PNP (positive negative positive), 59, 172
purpose of, 55
selecting ICs, 71
selecting transistors, 64, 176
stocking, 64
temperature-sensing ICs, 209, 212, 214
transistor pinouts, 412
TRIAC (TRIode for alternating current), 67
transmitters, legal aspects of, 349
(see also radio frequency)
TRIAC (TRIode for alternating current), 67
trimmers/trimpots, 16
tristate logic, 172
TTL (transistor transistor logic), 283
typographical conventions, xiv

U

unijunction transistors, 55
unipolar stepper motors, 243
units and prefixes, 423

V

V (volts), 3
VA (volt amps), 83
variable output voltage regulators, 89
variable resistors, 15, 211
VCO chips, 354
voltage
    AC conversion, 41, 47, 82
    avoiding drops from input to output, 299
    back-emf, 226

breakdown voltage (diodes), 49
calculating, 4
calculating within circuits, 7
comparing two levels of, 332
DC restrictions, 49
defined, 2
forward (Vf), 48
level conversion in microcontrollers, 165
measuring, 391
measuring AC, 392
measuring high, 397
reducing to measurable levels, 19
reference voltages, 49
reverse-biased, 48
ripple voltage, 84, 85
root mean square (RMS) voltage, 10
safety warning, 70, 95, 100, 406
switching very high, 63
voltage dividers, 19, 397
voltage ratings (capacitors), 35
voltage-controlled oscillators (VCO), 310
voltage-multiplier circuits, 98
volts (V), 3

W

watts (W), defined, 8
wires
    photos of, 26
    properties of common, 25
    selecting, 25
    standards for, 25

Z

Zener diodes, 49, 300